B A S I C
ASTRONOMY
L A B S

Jay S. Huebner

Michael D. Reynolds • Terry L. Smith

PRENTICE HALL Upper Saddle River, NJ 07458

Production Editor: *Kimberly Knox*
Production Supervisor: *Joan Eurell*
Acquisitions Editor: *Alison Reeves*
Assistant Editor: *Wendy Rivers*
Production Coordinator: *Ben Smith*

Printed in the United States of America

10 9 8 7 6 5 4 3

ISBN 0-13-376336-6

Prentice-Hall International (UK) Limited,London
Prentice-Hall of Australia Pty. Limited, Sydney
Prentice-Hall Canada Inc., Toronto
Prentice-Hall Hispanoamericana, S.A., Mexico
Prentice-Hall of India Private Limited, New Delhi
Prentice-Hall of Japan, Inc., Tokyo
Pearson Education Asia Pte. Ltd., Singapore
Editora Prentice-Hall do Brasil, Ltda., Rio de Janeiro

Table of Contents
Basic Astronomy Labs: A Laboratory Text
Huebner, Reynolds, Smith

Foreword

This manual has evolved from the exercises conducted in the basic astronomy laboratory course taught at the University of North Florida for the past 6 years. It contains 43 lab exercises that are appropriate for a one semester or year long laboratory course. Each exercise contains an introduction that should make it clear to beginning students why the particular topic of that lab is of interest and relevant to astronomy. No mathematics beyond simple high school algebra and trigonometry are required, and an exercise reviewing that is provided (Exercise 37, Astronomy Math Review). About one-third of the exercises are observing exercises, several of which can be repeated with different subjects, for example, Exercise 12, Astrophotography, and Exercise 29, Planetary Observing.

The exercises provided here include variations on standard and popular exercises, and also many exercises which many astronomy instructors will find to be new and innovative. Those which we believe to be innovative are listed below with brief comments. The exercises are organized in this manual into six major topics: Sky, Optics and Spectroscopy, Celestial Mechanics, Solar System, Stellar Properties, and Exploration and Other Topics.

The exercises we believe to be innovative are:

Exercise 5, The Messier List, a pen and paper exercise, which has students discover basic facts about the Milky Way Galaxy by plotting these objects on a star chart.

Exercise 6, About Your Eyes, a lab exercise, which has students measure the time required for their eyes to adapt to the dark. Students also learn to see the light which passes through their eye lids.

Exercise 12, Astrophotography, an observing exercise, which includes a discussion of how photography makes brighter stars appear larger.

Exercise 13, Electronic Imaging, an observing exercise, which explores modern electronic methods of recording images, especially the use of CCD's.

Exercise 15, Motions of Earth, a pen and paper exercise in which students discover just how fast the Earth is moving through space and in which direction is it going.

Exercise 18, Orbiting Earth, a pen and paper exercise in which students learn the basics of space transportation.

Exercise 23, Solar Observing, an observing exercise in which students generate data that they use in a subsequent exercise, to determine the rate of the Sun's rotation, verify the equation of time, and determine the latitude and longitude of the observing site.

Exercise 24, Solar Eclipses, an observing exercise in which students make quantitative and qualitative observations of solar eclipses.

Exercise 27, Lunar Eclipses, an observing exercise in which students make quantitative and qualitative observations of lunar eclipses.

Exercise 28, Observing Comets, an observing exercise for bright comets.

i

Exercise 30, Occultations, an observing exercise for lunar and planetary occultations.

Exercise 33, Elements and Supernovae, a pen and paper exercise in which students explore the periodic table to learn the properties of elements that are important for differentiation of the Solar System, and radioactive heating and dating of celestial bodies.

Exercise 35, Binary Stars, an observing exercise in which students make quantitative and qualitative observations of binary stars.

Exercise 36, Variable Stars, an observing exercise in which students make quantitative and qualitative observations of variable stars.

Exercise 38, Computer Planetaria, a computer exercise in which students explore these useful tools of astronomy.

Exercise 39, Astronomy on the Internet, a computer exercise in which students explore this enormous source of astronomical information.

Exercise 40, Observatory Visit, an exercise for directing a tour of an observatory.

Exercise 41, Planetarium Visit, an exercise for directing a visit to a planetarium.

Exercise 42, Radioactivity and Time, a lab exercise in which students measure the half-life of a short-lived isotope, and consider radioactive dating and heating of celestial bodies.

The following labs contain observing exercises:

Exercise 3, Sky Patterns
Exercise 4, Dark Sky Observing
Exercise 12, Astrophotography
Exercise 13, Electronic Imaging
Exercise 23, Solar Observing
Exercise 24, Solar Eclipses
Exercise 26, Lunar Observing
Exercise 27, Lunar Eclipses
Exercise 28, Observing Comets
Exercise 29, Planetary Observing
Exercise 30, Occultations
Exercise 35, Binary Stars
Exercise 36, Variable Stars

Guide to Astronomical Pronunciation

This is a list of rules for the pronunciation of astronomical names. Most of the astronomical names used today have their origins in Latin, Classical Greek, and Arabic. An effort has been made to preserve as many of the original sounds as possible. Because these languages were unplanned, some letters have more than one sound. A good rule for pronouncing a particular letter is to choose the sound that is easiest to say with the other sounds in the word.

Vowels

a: as in "father" or as in "barn"
e: as in "let" or as 'ay' in "play"
i: as in "machine" or as in "bit"
o: as in "note" or as in "knot"
u: as in "rude" or as in "put"
y: as in the German "über" or as 'i' in "bit."

Dipthongs

ae: as 'ai' in "aisle"
au: as 'ou' in "out"
ei: as in "reign"
eu: as "e(h)oo" or as in "neutral"
oe: as 'oi' in "oil"
ui: as in "ruin."

Consonants

Pronounce consonants as in American English with the following specifications:

c: as in "can"
g: as in "go"
i and **j**: as 'y' in "yet"
qu: as 'k' in "kite" or as in "quill"
r: same as in American English, except trilled
s: as in "sea"

t: as in "tired"
v: as 'w' in "wake" or as in "vase"
x: as in "axle"
ch: as in the German "Bach" or as 'ck' in "block"
ph: as in "philosophy"
th: as in "thin"

double consonants are *pronounced double*, that is, they are held longer than single consonants

Emphasis

Words have as many syllables as vowels. Normally, emphasis (accent) falls on the antepenult (last syllable but two). The penult (next to last syllable) gets the accent if there are only two syllables (as in Rigel), the penult contains a long vowel or a dipthong (as in Cassiopeia), or the penult's vowel is followed by two or more consecutive consonants (as in Sagitta).

Examples

Acamar - AK-a-mar
Aldebaran - ahl-DEB-ahr-ahn
Boötes - BOH-oh-tes
Betelgeuse - beh-tel-GEU-seh
Canes Venatici - KAHN-ehs weh-NAH-tee-kee
Canis Major - KAHN-ees MAH-yor
Cassiopeia - kas-see-o-PAY-a
Delphinus - DEHL-fee-nus
Enif - EH-nif
Equuleus - eh-KOO-leh-us
Gemini - GEH-mee-nee
Hyades - HOO-ah-des
Lacerta - lah-KER-ta
Lupus - LOO-pus

Mizar - MEE-zahr
Monoceros - mo-NO-kehr-os
Nunki - NOON-kee
Orion - O-ree-ohn
Pisces - PEES-kehs
Pleiades - PLEH-ah-des
Procyon - PRO-koo-on
Rigel - REE-gehl
Sagitta - sah-GEET-tah
Sagittarius - sah-geet-TAH-ree-us
Taurus - TOU-rus (ou in house)
Ursa Minor - OOR-sa MEE-nor
Vulpecula - wool-PEH-koo-lah
Zubenelgenubi - zoo-beh-nehl-GEH-noo-bee

Guide to the Constellations

Our ancestors gazed up at the night sky, without air and light pollution, and saw great beauty and mystery. They imagined pictures in the sky, formed by the tiny points of light known as stars. These pictures are the constellations. Each culture has had its own set of constellations, depicting the concerns of its people. The early European sailors to the southern hemisphere named the imaginary star patterns after concerns of their time (c.f., Vela (sails), Puppis (ship's stern), Sextans (sextant), Telescopium (telescope), etc.). Had the constellations been named in our modern age, we might have placed computers, digital watches, televisions, and mushroom clouds in the sky.

The constellations are named in Latin, which adds a little confusion but a lot of romance to the heavens. The constellations are also named in the international language, Esperanto. Modern star charts divide the heavens into eighty-eight constellations. Although celestial coordinates are more precise for locating objects, modern astronomers continue to use the constellations as a convenient way to communicate the general region of the sky in which to find an object.

The easiest way to find out which stars will be visible at a particular time and date is to use a star wheel. Just dial the star wheel to the desired date and time. The stars visible through the window are the ones in the sky at that time. For example, at 2100 (or 9:00 PM) on 14 February, Coma Berenices and Leo are rising in the east with Virgo not far behind. Ursa Major is in the northeast. Nearly overhead are Auriga, Gemini, and Cancer. To the south are Taurus, Orion, and Canis Major. In the west are Perseus, Cassiopeia, and Andromeda. In the northwest is Cepheus, and in the southwest is Cetus.

The mythology behind the names of the constellations illustrate the imaginations of some of our ancestors. Most people are familiar with the story of Perseus saving the fair Andromeda from the ravening sea-monster, Cetus. Andromeda's parents, Cepheus, the king and Cassiopeia, the queen, stand by helplessly. Other stories from many cultures are available at your local library or bookstore.

Along with constellations, other sky patterns also have been named. These patterns are called asterisms. An asterism is composed of part of a larger constellation or from parts of more than one constellation. For example, the Big Dipper is part of the larger constellation Ursa Major (The Greater Bear) and the Summer Triangle is composed of the brightest stars from the constellations Lyra, Cygnus, and Aquila. Other famous asterisms are the Little Dipper, the Great Square, the Seven Sisters, and the Winter Pentagon.

Sometimes the sky is shown as a globe with the stars pasted to its surface and Earth resting at its center. As Earth rotates about its axis from west to east, the sky seems to rotate from east to west. Stars move across the sky, some rising, some setting, and some neither rising nor setting. These stars that never fall below the horizon are called circumpolar stars, for they move in circular paths around the celestial poles. In the northern hemisphere, the point in the sky that does not seem to move at all is the north celestial pole. Currently, Polaris, the North Star, is the closest star visible with the naked eye to the north celestial pole. Polaris did not always hold this position.

Due to Earth's axial precession, the celestial poles move over time. When the great pyramids of Giza were being built, Thuban in the constellation of Draco held this position. In thirteen thousand years, Vega in Lyra will be near the north celestial pole. Vega shall make a fine north star, for it is the fifth brightest star in the night sky.

Most of the bright stars have names. But, astronomers need an easier way to label stars. The Bayer Constellation Designation (BCD) names a star based on the constellation in which it appears and its relative brightness compared with other stars in the constellation. The first part of the BCD is a Greek letter. The brightest star of the constellation is labeled Alpha, the second brightest is labeled Beta, and so on. The second part of the designation is the genitive form of the constellation name. Regulus, the brightest star in the constellation of Leo, is given the BCD of Alpha Leonis. The third brightest star in Crux is Gamma Crucis.

Unfortunately, brightnesses have been incorrectly measured or stars have changed their brightnesses since their BCD's were established. So, the BCD ordering may not always accurately reflect the brightnesses of the stars. For example, Betelgeuse is listed as Alpha Orionis, though it is dimmer than Rigel (Beta Orionis). Modern catalogues of stars label the stars with their celestial coordinates (right ascension and declination).

The following table lists for each of the eighty-eight constellations its Latin name, its genitive form, the Latin abbreviation, the Esperanto [1] name, and an English description.

Latin Name	Genitive	Abbr	Esperanto	English Description
Andromeda	Andromedae	And	Andromedo	Chained Princess
Antlia	Antliae	Ant	Pumpilo	Air Pump
Apus	Apodis	Aps	Birdo de Paradiso	Bird of Paradise
Aquarius	Aquarii	Aqr	Akvoportanto	Water Bearer
Aquila	Aquilae	Aql	Aglo	Eagle
Ara	Arae	Ara	Altaro	Altar
Aries	Arietis	Ari	Sxafo	Ram
Auriga	Aurigae	Aur	Cxargvidisto	Charioteer
Boötes	Boötis	Boö	Brutisto	Herdsman
Caelum	Caeli	Cae	Cxizilo	Sculptor's Chisel
Camelopardalis	Camelopardalis	Cam	Gxirafo	Giraffe
Cancer	Cancri	Cnc	Kankro	Crab
Canes Venatici	Canum Venaticorum	CVn	Cxasantoj Hundoj	Hunting Dogs
Canis Major	Canis Majoris	CMa	Hundego	Greater Dog
Canis Minor	Canis Minoris	CMi	Hundeto	Lesser Dog
Capricornus	Capricorni	Cap	Kaprikorno	Sea Goat
Carina	Carinae	Car	Kilo	Ship's Keel
Cassiopeia	Cassiopeiae	Cas	Kasiopejo	Queen in a Chair
Centaurus	Centauri	Cen	Centauro	Centaur
Cepheus	Cephei	Cep	Kefeo (Cefeo)	Monarch
Cetus	Ceti	Cet	Baleno (Ceto)	Whale

Chamaeleon	Chamaeleontis	Cha	Kameleono	Chameleon
Circinus	Circini	Cir	Cirkelo	Pair of Compasses
Columba	Columbae	Col	Kolombo	Dove
Coma Berenices	Comae Berenices	Com	Haroj de Berenico	Berenice's Hair
Corona Australis	Coronae Australis	CrA	Krono Suda	Southern Crown
Corona Borealis	Coronae Borealis	CrB	Krono Norda	Northern Crown
Corvus	Corvi	Cor	Korvo	Crow
Crater	Crateris	Crt	Kratero (Taso)	Cup
Crux	Crucis	Cru	Kruco	Cross (Southern)
Cygnus	Cygni	Cyg	Cigno	Swan (Northern Cross)
Delphinus	Delphini	Del	Delfeno	Dolphin
Dorado	Doradus	Dor	Glavfisxo	Swordfish
Draco	Draconis	Dra	Drako	Dragon
Equuleus	Equulei	Equ	Cxevaleto	Little Horse
Eridanus	Eridani	Eri	Eridano	River Eridanus (Po)
Fornax	Fornacis	For	Forno	Furnace
Gemini	Geminorum	Gem	Gxemeloj	Twins
Grus	Gruis	Gru	Gruo	Crane
Hercules	Herculis	Her	Herkulo	Hercules, son of Zeus
Horologium	Horologii	Hor	Horologxo	Clock
Hydra	Hydrae	Hya	Hidro	Sea Serpent
Hydrus	Hydri	Hyi	Marserpento	Water Snake
Indus	Indi	Ind	Hindo	Indian
Lacerta	Lacertae	Lac	Lacerto	Lizard
Leo	Leonis	Leo	Leono	Lion
Leo Minor	Leo Minoris	LMi	Leoneto	Lesser Lion
Lepus	Leporis	Lep	Leporo	Hare
Libra	Librae	Lib	Skalo	Balance (Scales)
Lupus	Lupi	Lup	Lupo	Wolf
Lynx	Lyncis	Lyn	Linko	Lynx
Lyra	Lyrae	Lyr	Liro	Lyre (Harp)
Mensa	Mensae	Men	Tablo	Table Mountain
Microscopium	Microscopii	Mic	Mikroskopo	Microscope
Monoceros	Monocerotis	Mon	Unukornulo	Unicorn
Musca	Muscae	Mus	Musxo	Fly
Norma	Normae	Nor	Nivelilo	Carpenter's Level
Octans	Octantis	Oct	Oktanto	Octant
Ophiuchus	Ophiuchi	Oph	Ofiuhxo (Ofiuko)	Ophiuchus, the Serpent Bearer
Orion	Orionis	Ori	Oriono	Orion, the Hunter
Pavo	Pavonis	Pav	Pavo	Peacock

Pegasus	Pegasi	Peg	Pegazo	Pegasus, the Winged Horse
Perseus	Persei	Per	Perseo	Perseus, the Hero
Phoenix	Phoenicis	Phe	Fenikso	Phoenix
Pictor	Pictoris	Pic	Stablo	Painter's Easel
Pisces	Piscium	Psc	Fisxoj	Fishes
Piscis Austrinus	Piscis Austrini	PsA	Fisxo Suda	Southern Fish
Puppis	Puppis	Pup	Poupo	Ship's Stern
Pyxis	Pyxidis	Pyx	Kompaso	Mariner's Compass
Reticulum	Reticuli	Ret	Reto	Net
Sagitta	Sagittae	Sge	Sago	Arrow
Sagittarius	Sagittarii	Sgr	Pafarkisto	Archer
Scorpius	Scorpii	Sco	Skorpio	Scorpion
Sculptor	Sculptoris	Scl	Skulptisto	Sculptor
Scutum	Scuti	Sct	Sxildo	Shield
Serpens	Serpentis	Ser	Serpento	Serpent
Sextans	Sextantis	Sex	Sekstanto	Sextant
Taurus	Tauri	Tau	Tauro	Bull
Telescopium	Telescopii	Tel	Teleskopo	Telescope
Triangulum	Trianguli	Tri	Triangulo	Triangle
Triangulum Australe	Trianguli Australis	TrA	Triangulo Suda	Southern Triangle
Tucana	Tucanae	Tuc	Tukano	Toucan
Ursa Major	Ursae Majoris	UMa	Ursego	Greater Bear
Ursa Minor	Ursae Minoris	UMi	Urseto	Lesser Bear
Vela	Velorum	Vel	Velo	Ship's Sails
Virgo	Virginis	Vir	Virgo	Maiden
Volans	Volantis	Vol	Fisxfluganto	Flying Fish
Vulpecula	Vulpeculae	Vul	Vulpeto	Little Fox

[1] Esperanto is a planned language developed by L.L. Zamenhoff in the late 19th century. The alphabet is pronounced similar to American English, except each letter has one sound only; i.e., a as in father, b as in bear, c as 'ts' in bats, d as in dog, e as in terran, f as in fox, g as in garden, h as in hot, i as in machine, j as 'y' in year, k as in kite, l as in light, m as in moon, n as in noon, o as in orbit, p as in Pluto, r as in ray (trilled), s as in sail, t as in torque, u as in rule, v as in vector, z as in zodiac. Consonants followed by an 'x' are softened; i.e., cx = 'ch' in charge, gx = 'g' in gem, hx = 'ch' in Bach, jx = 's' in pleasure, sx = 'sh' in ship. Also, the following dipthongs are available: au = 'ou' in house, aj = 'i' in light, ej = 'ei' in vein, oj = 'oy' in joy, uj = 'ui' in ruin. A word has as many syllables as vowels or dipthongs. Emphasis always falls on the penultimate (next-to-last) syllable.

The International System of Units

Le Système International d'Unités (SI), the International System of Units, is the modern metric system. It is a standard system of measurements established by international agreement. SI has been used in the United States since 1866. In fact, the yard and pound have been based on the meter and the kilogram since 1893. In the 1950's, the inch was redefined to be exactly 2.54 centimeters.

The three most often used base units of SI are the meter (length), the kilogram (mass), and the second (time). Hence, the system is often called the **MKS system**. Other base units are the kelvin (absolute temperature), the ampere (electric current), the mole (amount of a substance), and the candela (luminous intensity). Some supplementary units are the radian (plane angle measure) and the steradian (solid angle measure). Some older texts may use the cgs (centimeter -gram- second) system, but it is a simple matter to convert between these units and the standard MKS units.

All other SI units, called "derived units," are defined in terms of these units. Table 1 contains a listing of some common SI units and their abbreviations. For example, the newton (N) is the SI unit of force and is defined as the force required to accelerate a 1 kilogram mass at a rate of 1 meter per second per second. Another common derived unit is the liter (L), a unit of volume. The liter is equivalent to 1000 cm^3, so the milliliter (mL) is used interchangeably with the cubic centimeter (cm^3 or cc).

Multiples and fractions of these units may be made using standard prefixes. Table 2 lists the standard prefixes and their values. Notice that these multiples and fractions are based on powers of ten. This makes for easy conversion between metric units. For example, 34.5 kilometers is 34,500 meters or 3.45 million centimeters.

Of the industrialized countries, only the United States of America has not completely converted to the metric system. This is causing an increasing number of problems in trade. Some of the archaic U.S. Customary units are still in use and are listed in Table 3 with factors for converting them to SI units. For example, 10 acres is about 4 hectares.

Table 1: Some SI Units

Unit Name	Abbreviation	Derivation	Description
meter	m		length
kilogram	kg		mass
second	s or sec		time
kelvin	K		absolute temperature
ampere or amp	A		electric current
mole	mol		amount of a substance
candela	cd		luminous intensity
radian	rad		plane angle measure
steradian	sr		solid angle measure
degree Celsius	°C	$(K - 273.15) \cdot (1°C/K)$	temperature
metric ton (tonne)	ton or tonne	1000 kg	mass
are	are	100 m^2	area
liter	L	1000 cm^3	volume
hertz	Hz	$1/s$	frequency
newton	N	$kg \cdot m/s^2$	force
joule	J	$N \cdot m$	energy, work, heat
watt	W	J/s	power
pascal	Pa	N/m^2	pressure
volt	V	W/A	electric potential
weber	Wb	$V \cdot s$	magnetic flux

Table 2: Standard Metric Prefixes

Prefix	Abbreviation	Value	Prefix	Abbreviation	Value
exa	E	$\cdot 10^{18}$	deci	d	$\cdot 10^{-1}$
peta	P	$\cdot 10^{15}$	centi	c	$\cdot 10^{-2}$
tera	T	$\cdot 10^{12}$	milli	m	$\cdot 10^{-3}$
giga	G	$\cdot 10^{9}$	micro	m	$\cdot 10^{-6}$
mega	M	$\cdot 10^{6}$	nano	n	$\cdot 10^{-9}$
kilo	k	$\cdot 10^{3}$	pico	p	$\cdot 10^{-12}$
hecto	h	$\cdot 10^{2}$	femto	f	$\cdot 10^{-15}$
deka	da	$\cdot 10^{1}$	atto	a	$\cdot 10^{-18}$

Table 3: Some Archaic Units and Conversion Factors to SI Units

start with	multiply by	get
inch	2.54	centimeter
foot	0.3048	meter
cubit	0.4826	meter
fathom	1.8288	meter
rod	5.0292	meter
mile	1.609	kilometer
slug	14.59	kilogram
pound (mass)	0.4536	kilogram
pound (force)	4.448	newton
U.S. short ton	907.2	kilogram
British thermal unit (BTU)	1,055	joule
acre	0.4047	hectare
U.S. gallon	3.7861	liter
pound per square inch (psi)	6,895	Pascal

Hand-Angle Measurements

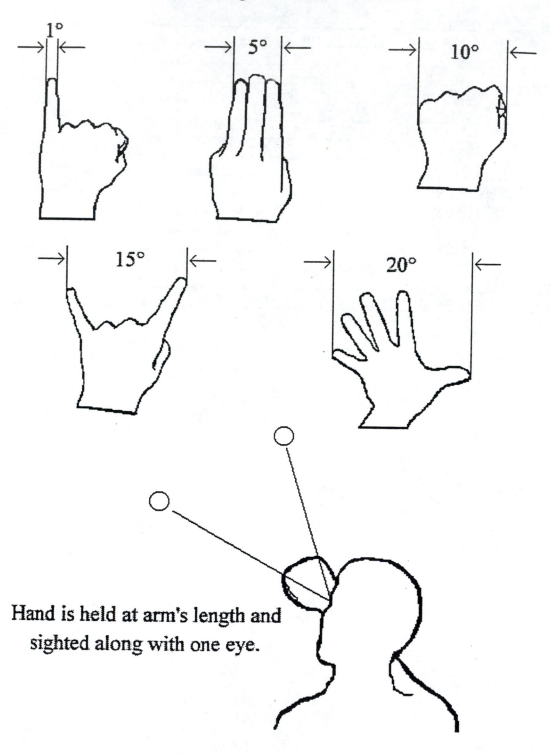

Hand is held at arm's length and sighted along with one eye.

Spectral Classifications

In the late 1800's and early 1900's, astronomers noticed that stars could be grouped according to their spectra. Edward C. Pickering and Annie J. Cannon developed a method of grouping stars with similar spectral characteristics. This spectral classification scheme is still used today, though it has been expanded somewhat. See Exercise 32, Hertzsprung-Russell Diagram, for some exercises using spectral classes.

Pickering and Cannon originally classified stars according to the number and intensity of the hydrogen absorption lines in their spectra. Stars with the strongest hydrogen lines were called type A, stars with the second strongest hydrogen lines were called type B, and so on. By 1920, the spectra of several hundred thousand stars from both the northern and southern hemispheres had been catalogued. Table 1 lists some properties of the stars in each spectral class.

The 1920's saw the emergence of our modern understanding of atomic structure. See Exercise 11, Atomic Spectra, for more information. Analysis of stellar spectra by astronomers, like Meghnad N. Saha and Cecilia Payne-Gaposchkin, led to the realization that the spectral differences were due to differences in the surface temperatures of the stars. The spectral classes were resorted according to temperature and redundant classes were removed. In the current arrangement, the spectral classes, running from hottest to coolest, are **W, O, B, A, F, G, K, M, R, N,** and **S**. There are several popular mnemonics for remembering this sequence. One is "Wow! Oven Baked Ants, Fried Gently, Kept Moist, Retain Natural Succulence."

The first and last three spectral classes were added more recently than the others. Type W stars (also called Wolf-Rayet stars) are fairly young stars whose outer atmospheres have been blown away by strong stellar winds and, perhaps, by gravitational interaction with companion stars. The spectra of Wolf-Rayet stars differ from those of other classes by exhibiting only emission lines. Stars of types R, N, and S are called carbon stars, for they show absorption lines by carbon molecules. Type S stars seem to be between M- and N-type stars, but with zirconium oxide (ZrO) lines instead of titanium oxide (TiO) lines.

Annie J. Cannon recognized that the simple lettering system was too crude to fully describe the subtleties of the observed spectra. She divided each spectral class into ten temperature subdivisions, numbered 0 through 9. The hottest stars of a given class are assigned 0, the second hottest are assigned 1, and so on through 9.

Finally, a luminosity class is used to note the position of the star on the HR diagram. See Exercise 32, Hertzsprung-Russell Diagram. These classes are Ia (bright supergiants), Ib (supergiants), II (bright giants), III (giants), IV (subgiants), V (main sequence stars/dwarfs), VI (white dwarfs), and VII (subdwarfs). For example, the Sun is classified as G2V, meaning it has a spectral class of G, it is in the third hottest group of that class, and it is on the main sequence. Rigel (Beta Orionis) is a B8Ia star, meaning it is in the ninth hottest group of the B spectral class, and it is a bright supergiant.

Table 1: Some Properties of the Spectral Classes

Class	Characteristics[1]	Approx. Color	Surface Temperature	Examples
W	resemble type O stars but with broad emission features due to their turbulent atmospheres	blue	50,000 - higher	γ Vel
O	ionized He, N, and O; weak H	bluish	28,000 - 50,000	χ Per, λ Ori
B	ionized H, neutral He	blue-white	9,900 - 28,000	Rigel, Regulus
A	strong H, ionized Mg, Si, Fe, Ti, Ca, etc.	white	7,400 - 9,900	Sirius, Altair
F	weaker H, ionized Fe, Cr, Ti, and other metals; some neutral metals	yellow-white	6,000 - 7,400	Canopus, Procyon
G	ionized Ca, ionized and neutral metals	yellow-white	4,900 - 6,000	Sol, α Cen A
K	ionized Ca, hydrocarbon molecules, neutral metals	orange	3,500 - 4,900	Arcturus, Pollux
M	TiO, singly ionized Ca, neutral metals, other molecules	reddish	2,000 - 3,500	Antares, Betelgeuse
R	CN, weak C bands	orange	3,500 - 5,400	S Cam
N	C_2, other carbon molecules	red	1,900 - 3,500	R Lep, S Cep
S	ZrO, other molecules, H emission lines	red	2,000 - 3,500	R Cyg

1. Chemical symbols from the periodic table of the elements are used in this column. See Exercise 33, Elements and Supernovae, for a periodic table.

Quotes on Science
or knowing the Universe through a few sentences

The contemplation of celestial things will make a man both speak and think more sublimely and magnificently when he descends to human affairs.

Marcus Tullius Cicero, 106-43 B.C.

. . . each one of us and all of us, are truly and literally a little bit of stardust.

W. A. Fowler, 1911-

The mind, once expanded to the dimensions of larger ideas never returns to its original size.

Oliver Wendell Holmes, 1809-1894

Knowledge is power.

Francis Bacon, 1561-1626

We know very little, and yet it is astonishing that we know so much, and still more astonishing that so little knowledge can give us so much power.

Bertrand Russell, 1872-1970

The whole of science is nothing more than the refinement of everyday thinking.

Albert Einstein, 1879-1955

It has been said that there is no such thing as a free lunch. But the universe is the ultimate free lunch.

Alan Guth, 1947-

The diversity of the phenomena of nature is so great, and the treasures hidden in the heavens so rich, precisely in order that the human mind shall never be lacking in fresh nourishment.

Johannes Kepler, 1571-1630

In questions of science, the authority of a thousand is not worth the sound reasoning of a single individual.

Galileo Galilei, 1564-1642

The task is not so much to see what no one has yet seen, but to think what nobody has yet thought about that which everybody sees.

Erwin Schroedinger, 1887-1961

People see only what they are prepared to see.

Ralph Waldo Emerson, 1803-1882

We do not see things as they are, we see things as we are.

The Talmud, 1532

Chance favors the prepared mind.

L. Pasteur, 1822-1895

Everything should be as simple as possible, but not simpler.

Albert Einstein, 1879-1955

The house of delusion is cheap to build, but drafty to live in.

A. Housman, 1859-1936

Science, measured against reality, is primitive and childlike - and yet it is the most precious thing we have.

Albert Einstein, 1879-1955

We may now be near the end of the search for the ultimate laws of nature.

Stephen Hawking, 1942-

A scientist is a person who can find out things that nobody else can tell whether he found them out or not. And the more things he can find out that no one else can tell about, why the bigger scientist he is.

Will Rogers, 1879-1935

Space is what keeps everything from happening at the same place.

Jay Huebner, 1939-

Space acts on matter telling it how to move. In turn, matter reacts back on space telling it how to curve.

Charles Misner, 1932-

How inappropriate to call this planet Earth, when clearly it is Ocean.

A. C. Clark, 1917-

Our loyalties are to the species and the planet. Our obligation to survive is owed not just to ourselves, but to that cosmos, ancient and vast, from which we sprang.

Carl Sagan, 1934-

One important, often overlooked fact about the solar system is that the bulk of the real estate is not on planets.

Freeman J. Dyson, 1923-

Time is what prevents everything from happening at once.

John Wheeler, 1911-

... in the beginning was the plasma.

Hannes Alfven, 1908-

Prehistory ended with the Big Bang.

Jay Huebner, 1939-

The world was made, not in time, but simultaneously with time. There was no time before the world.

St. Augustine, 354-430

The world was created on 22 October 4004 BC at 6 o'clock in the evening.

James Ussher, 1581-1656

Space and time are not conditions in which we live, but modes in which we think.

Albert Einstein, 1879-1955

To try to write a grand cosmical drama leads necessarily to myth.

Hannes Alfven, 1908-

Common sense is a particular group of prejudices acquired before the age of 18.

Albert Einstein, 1879-1955

The most incomprehensible thing about the universe is that the universe is so comprehensible.

Albert Einstein, 1879-1955

The universe is not only queerer than we suppose, it is queerer than we can suppose.

J. B. S. Haldane, 1892-1964

The more the universe seems comprehensible, the more it also seems pointless.

Steven Weinberg, 1933-

To many the search for life in the universe is the greatest adventure left to humanity.

Frank Drake, 1930-

Either we are alone in the Universe, or we are not. In either case, it is amazing.

L. Dubridge, 1901-

Geometry provided God with a model for the creation.

Johannes Kepler, 1571-1630

Penetrating so many secrets, we cease to believe in the unknowable. But there it sits never-the-less, calmly licking its chops.

H. L. Mencken, 1880-1956

Why shouldn't truth be stranger than fiction, after all fiction has to make sense.

Mark Twain (S. Clemens), 1835-1910

Science is organized knowledge.

H. Spencer, 1820-1903

Science has to do with ordering complexity.

Jay Huebner, 1939-

Knowledge is one. Its division into (academic) subjects is a concession to human weakness.

H. J. Mackinder, 1861-1947

When I want to read a good book, I write one.

B. Disraeli, 1804-1881

Lots of things are invisible, but we don't know how many because we can't see them.

Dennis the Menace, created by Hank Ketcham, 1920-

Observer:
Exercise:
Date:
Lab Day:

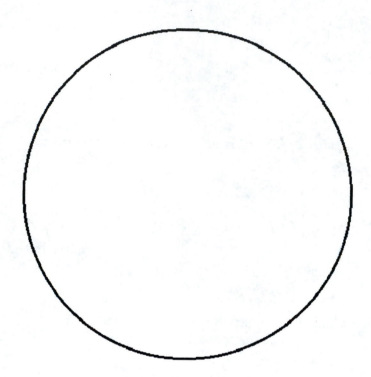

Observing Form B

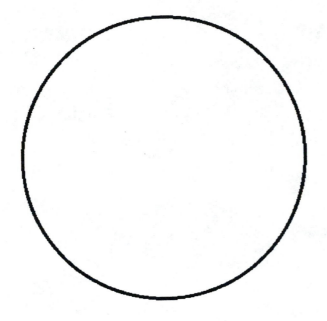

Notes:

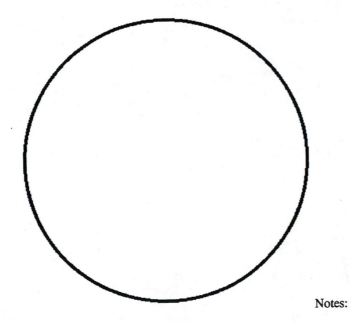

Notes:

Observing Form C

Observer:
Exercise:
Date:
Lab Day:

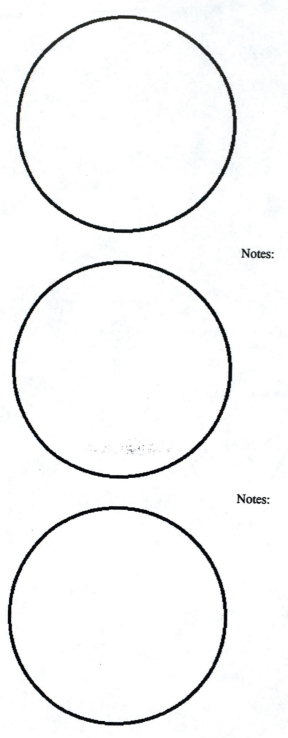

Notes:

Notes:

Notes:

Star Wheels

Our ancestors watched the Sun, Moon, and planets move against the background of stars. They learned to tell time and predict the changing of the seasons by the heavens. The modern observer, too, will find it useful to understand the motions in the heavens. The star wheel is a simple device and has advantages over star charts in that it can show how the stars move as well as where they are for any date and time.

Most star wheels have two parts, the wheel and the jacket, and are designed for a specific latitude. The wheel is a circular surface, mounted so that it may turn within the jacket. The wheel has star patterns printed on it with date marks along its circumference. The jacket has hour marks printed along its edge and an oval hole cut in it so part of the wheel is visible through the jacket. When a date mark on the wheel is aligned with an hour mark on the jacket, the star patterns visible at that date and time are seen in the oval hole. The oval hole, thus, represents the sky and its perimeter represents the horizons of the observer. The star wheel should be held overhead with the north end pointing north, so it can be viewed from below as the sky is viewed. In this way, the star wheel simulates the dome of the sky.

It is sometimes useful to imagine the sky objects fixed to a celestial sphere with Earth at its center. Relative to Earth, this celestial sphere rotates to the west. The celestial equator is a projection of Earth's equator onto this celestial sphere. So, the celestial equator passes overhead for an observer on Earth's equator. From the north pole of Earth, the point at the zenith is the north celestial pole. Currently, a medium-bright star, Polaris, is near the north celestial pole; this is the so-called North Star. There is no bright star near the south celestial pole.

Any point on the celestial sphere can be specified with celestial coordinates, **right ascension (RA) and declination (DEC)**. These are equivalent to the longitude and latitude, respectively, of the celestial sphere. Declination is the angle measure north (+) or south (-) of the celestial equator. DEC ranges from -90° at the south celestial pole to 0° at the celestial equator to +90° at the north celestial pole.

Right ascension is the measure eastward from the celestial prime meridian. This reference for RA is the great semi-circle that includes the celestial poles and the position of the Sun on the vernal equinox. On the star wheel, this is a line drawn from the celestial pole to the date mark of 21 March. RA is measured in hours (0^h through 23^h) around the celestial sphere. Each hour of RA is 15° along the celestial sphere from the next hour.

Not all stars can be seen from a given latitude, since Earth is in the way and blocks half of the celestial sphere from view. Figure 1-1 illustrates this for 30° N latitude. At this latitude, the north celestial pole is 30° above the northern horizon and the celestial equator is 30° south of the zenith. For an observer at 30° N latitude, stars within 30° of the north celestial pole never set below the horizon. These stars are called **circumpolar** and are always in the sky, though the sky is often too bright in the daytime to notice them. The south celestial pole is 30° below the southern horizon. Any star within 30° of the south celestial pole never rise above the horizon. Star wheels are made for specific latitudes and show only those stars that are above the horizon at the set date and time.

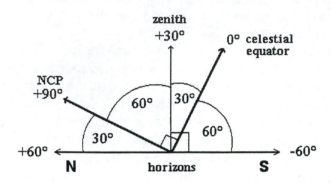

Figure 1-1: The angles between the horizons, zenith, north celestial pole, and equator for 30° north latitude.

A star wheel can be used to determine approximate celestial coordinates. A line drawn from the celestial pole to the vernal equinox date (roughly, 21 March) indicates RA 0^h, the celestial prime meridian. Every 15° eastward of this line is another hour of right ascension. If 21 March is aligned with an hour mark on the jacket, one can use the hour marks to determine right ascension since each hour mark is 15° from the next. Declination can be measured on the star wheel if the scale is known. One way to determine the scale is to measure the distance from the celestial equator to the celestial pole. This distance is equivalent to 90° on the celestial sphere.

Example 1: The shortest distance between the north celestial pole and the northern horizon is 20. mm on a particular star wheel made for 40° N latitude. Therefore, the scale of this star wheel is 2° for every millimeter. So, if the distance between two stars on this wheel is 15.25 millimeters, the two stars will be separated by 30°30' in the sky.

Example 2: Use a star wheel to determine the celestial coordinates of Sirius (Alpha Canis Majoris).
    Step 1: Find Sirius and align the 21 March date mark with an hour mark so the star can be seen. In this case, the 8 p.m. mark will do.
    Step 2: Align a straight-edge from the celestial pole through Sirius to the jacket. In this example, the straight-edge should lie at about the 1:15 p.m. mark. Count the number of hours and estimate the minutes between the 21 March mark and

the straight-edge. Remember to count eastward (this is clockwise on most star wheels). The RA of Sirius is 06^h45^m.

Step 3: Measure the distance between the celestial equator and Sirius. The ruler should be aligned so that it passes through the north celestial pole, the celestial equator, and Sirius. For the star wheel of Example 1, this distance is about 8.4 mm.

Step 4: Use the scale of the star wheel to determine the DEC of Sirius. The DEC of Sirius is -16°43'. Note that DEC is negative if the object is south of the celestial equator.

Thus, the RA/DEC of Sirius has been determined to be 06^h45^m/-16°43'.

Some star wheels show the annual path of the Sun in the sky. This path is called the ecliptic. The circle marking this path is centered in the constellation of Draco, at about 18^h00^m/+66.5°, about 23.5° from the north celestial pole. This difference is caused by the tilt of Earth's axis to the axis of its orbit. The ecliptic and the celestial equator intersect at two points, in Pisces and in Virgo. The Sun reaches these intersections at the vernal and autumnal equinoxes, respectively. Since the Sun is always on the ecliptic, a star wheel that shows the ecliptic can be used to determine the position of the Sun on the celestial sphere. The orbits of the major planets in the Solar System lie nearly in the same plane as the Earth's orbit. Hence, the major planets can always be found near the ecliptic.

Example 3: In which constellation is the Sun on 01 August?

Rotate the star wheel so that the 01 August mark is aligned with the noon mark. The Sun is on the meridian at solar noon. Since the star wheel uses mean solar time, solar noon occurs at 12 p.m. on the star wheel. The Sun is where the ecliptic crosses the meridian (the line the runs from due north to due south). This is in the constellation of Cancer. If you check the dates associated with "signs" in horoscopes, you will find some disagreements. This arises from precession of Earth's axis. The workings of astrology were established over two thousand years ago, before precession was understood.

Procedures

Apparatus

star wheel, millimeter ruler, and drawing compass.

A. The Scale of the Star Wheel

1. The latitude for which your star wheel was made should be printed somewhere on it. Find and record this value.

2. Measure the distance between the celestial pole and the celestial equator using a millimeter ruler. If your star wheel does not show the celestial equator, the west-most star in the belt of Orion (known as Mintaka) has a declination of only -00°18' and may be used to approximately mark the celestial equator. A drawing compass can be used to draw a circle centered on the north celestial pole to represent the celestial equator on the star wheel.

3. Compute the scale of your star wheel in degrees per millimeter.

4. Measure the distance between the celestial pole and the point on the horizon nearest the pole. For northern hemisphere star wheels, this horizon should be due north. Use the scale of your star wheel to determine the angle represented by this distance. This angle should equal the latitude for which the star wheel is made. Compute the percentage error between your measurement and the latitude of the star wheel.

5. Construct an angle ruler for your star wheel. Determine how many millimeters are equivalent to 10°. Make ten marks on a piece of paper or thin cardboard so that each mark is 10° from the next in the scale of the star wheel. Label these marks 0° through 90°. This angle ruler can be used to measure angles on the star wheel.

6. The angular distance between Betelgeuse (Alpha Orionis) and Rigel (Beta Orionis) is 18°36'. Betelgeuse is at $05^h55^m/+07°24'$ and Rigel is at $05^h15^m/-08°12'$. Use your scale to determine the angle between these two stars. Compute the percentage error between your measurement and the true angular distance.

B. Circumpolar Stars

1. Use a drawing compass to draw a circle with a radius equivalent to your latitude from the celestial pole on your star wheel. All the stars within this circle are circumpolar at your latitude.

2. List some constellations that are, at least in part, circumpolar from your latitude.

3. What constellations are circumpolar from a latitude 30° closer to the pole than your latitude?

4. If your star wheel does not show the celestial equator, draw it on your star wheel using a drawing compass.

C. Celestial Coordinates on the Star Wheel

1. Draw a line from the celestial pole to the 21 March date mark on the edge of the wheel. Any star on this line has a right ascension of 0^h. This line is the celestial prime meridian.

2. Determine the celestial coordinates (RA/DEC) of the following stars: Aldebaran (Alpha Tauri), Regulus (Alpha Leonis), Spica (Alpha Virginis), and Antares (Alpha Scorpii).

3. At 2200 on 04 July 1776 CE, Saturn was at 12^h58^m/-03°31'. In which constellation did Saturn reside?

D. The Ecliptic

1. If your star wheel does not show the ecliptic, then use a compass to draw it on your star wheel. Place the pivot of the compass at 18^h00^m/+66.5° and sweep out a circle with a radius equivalent to 90°. On what two dates does the ecliptic cross the celestial equator?

2. The zodiacal constellations lie on or near the ecliptic. Begin at RA 0^h, list the constellations of the zodiac in order of increasing RA.

3. Determine the approximate dates during which the Sun resides in each zodiacal constellation.

4. What is the approximate RA/DEC of the Sun on 04 July?

E. Practice with the Star Wheel and Almanac

1. Complete the following table to determine the accuracy of the star wheel. Note: ignore longitude correction and Equation of Time for this problem.

Event	Date	Time (almanac)	Time (star wheel)	Difference (min.)
Sirius transits		0000		
Vega rises	1 April			
Deneb transits	27 September			
Sun rises	21 March	0600		
Sun sets	21 September			

2. The best time make observations of objects other than the Sun and Moon is when the Moon is not in the sky. Around the new phase, the Moon rises and sets at about the time

the Sun does, so it is out of the sky most of the night. List the date of the new Moon of every month this year.

3. Determine the RA/DEC of the brightest star of each of the twelve zodiacal constellations. Begin with Pisces and proceed to increasing RA. Determine the months during which Sun resides in each of the zodiacal constellations. Arrange all information in a neat table.

4. In which zodiacal constellation is the Sun on the following special dates: vernal equinox, summer solstice, autumnal equinox and winter solstice?

5. In which constellation does the Sun reside when it is at RA 15h? What is its declination? On what date does the Sun reach this position?

6. The times given in the almanac are standardized for the defining longitude of the time zones. So, you must correct for your longitude. For example, Deneb transits at midnight on 2 August, according to the almanac. However, if your longitude is 81.5° W, you will see Deneb transit at 0026 EST. Choose a planet and determine the rise times of this planet every 30 days for three months. Be sure to identify the dates you use.

7. Determine the dates of midnight transit of all planets less than 10 AU from the Sun for which midnight transit is listed on the almanac. Do NOT correct these values for your location, for both the almanac and the star wheel use standardized time. Use these dates on the star wheel to determine the RA/DEC of these planets. Unlike the stars, the RA/DEC for planets change throughout the year.

8. Locate the constellation in which M57 (NGC 6720) can be found. M57 is at $18^h51.7^m/+32°58'$.

Coordinate Systems

To be useful in astronomy, coordinate systems must be able to unambiguously specify the location of each point on the celestial sphere or on a celestial body. Three systems will be discussed. First, the Latitude/Longitude system is useful for specifying points on the surface of spherical bodies. We use this system on Earth, Moon, Mars, and many other bodies. Second, the Altitude/Azimuth system is useful for giving directions while observing, and for aiming some telescopes. Third, the Right Ascension/Declination system is used to describe the positions of objects on the celestial sphere, independent of the horizons or the observer's position. This system is used in aiming most telescopes with clock drives, which allow the telescopes to track objects in the sky as Earth rotates. Although this laboratory exercise does not discuss it, a fourth coordinate system is Galactic Altitude/Azimuth (or Galactic Latitude/Longitude). It is used to specify positions of objects in and around the Milky Way galaxy.

Latitude/Longitude. Any point on the surface of a spherical body can be specified by two angle coordinates. For rotating bodies, it is customary to orient the coordinate system with the axis of rotation. The latitude of an object is the measure of the minimum angle between a ray from the center of the body to the object on the surface and a ray from the center of the body to the equator. Thus, an object on the equator of the spherical body has a latitude of 0°. An object at the north pole of the body has a latitude of 90° N, and an object at the south pole of the body has a latitude of 90° S. We say that the reference for latitude is the equator. Lines of constant latitude run parallel to the equator. Lines of constant longitude run perpendicular to the lines of latitude and pass through the poles. The reference of longitude is arbitrary. On Earth, 0° longitude is called the Prime Meridian and consists of a great semi-circle that passes through Greenwich, UK, the location of the British Royal Astronomical Observatory. On the Moon, the prime meridian is the line of longitude closest to Earth (since one side of the Moon always points towards Earth, 0° lunar longitude is a line that splits the full Moon in half). The coordinates of Greenwich, UK are 51°28'38.2"N/0°00'00", which means Greenwich is 51 degrees, 28 minutes-of-arc, and 38.2 seconds-of-arc north of the equator and 0 degrees, 0 minutes-of-arc, and 0 seconds-of-arc from the Prime Meridian. Washington D.C. is located at about 38°54'N/77°01'W. Note: 1 degree = 60 minutes-of-arc = 3600 seconds-of-arc or 1° = 60' = 3600".

Time zones. An issue related to longitude (and also right ascension, to be discussed later) is how time is kept at various places on Earth. Since it takes about 24 hours for Earth to complete one rotation, astronomers typically use a 24-hour clock. The day begins at 0000 hours (midnight) and progresses as on a normal clock to 1200 (noon). After noon, the 24-hour clock continues counting, 1300 (1 PM), 1400 (2 PM), and so on. It would be silly for everyone to use the same time on their clocks. Consider what would happen if we all agreed to use the time on the clock in Greenwich, UK. At 1200 on the summer solstice in Greenwich, the Sun would be at its highest in the sky. However, at the same instant, the Sun would be west of the meridian (setting) for people in Siberia, it would be east of the meridian (rising) for people in Central America, and it would not be visible for people in

Hawaii. Consequently, Earth has been divided into 24 time zones that roughly follow the lines of longitude (spaced every 15°). Within a time zone, clocks are set to the time measured at the middle longitude of the time zone. Universal Time (UT) is kept at Greenwich, UK. This is our temporal reference point. Since many astronomical events have nothing to do with Earth's rotation (e.g., when a meteorite impacts the Moon), most astronomical times are recorded as UT. It is then up to the astronomer to convert to the time seen on his/her own clock. The reference longitude for eastern standard time (EST) is 75°W. This makes EST 5 hours earlier than UT, because 75°/15° per hour = 5 hours. Thus, if an astronomer in Greenwich sees a meteorite hit the Moon at 2300 UT, the clock of an astronomer in the eastern standard time zone, who saw the same event, would read 1800 EST. How events translate across time zones depends on the type of event being considered. In the previous example, the impact of a meteorite does not depend on the longitude of the observers, even though the times on the clocks are different. Some events, however, do depend on where the observer is located. For example, if a star transits the meridian at 2100 for someone located at 0° longitude, it will transit five hours later when the clock of someone located at 75°W longitude also reads 2100. But, since this event is dependent upon the rotation of Earth, everyone west of a time zone line will have to wait some additional time for the event to take place. In Washington D.C. (77°01'W), that same star will cross the meridian at about 2108 EST. Any event that depends on Earth's rotation (rise times, meridian transits, set times, etc.) must include this extra amount of rotation time, but, otherwise, requires no conversion from UT.

Altitude/Azimuth. In this system, two angles need to be specified: an angle to indicate direction in the horizontal plane (i.e., toward which horizon) and an angle to indicate how far above or below the horizon in that direction. Azimuth (or AZ) must cover an entire circle, so its values range from 0° to 360°. 0° represents due north, 90° is due east, 180° is due south, and 270° is due west. These angles and any angle in between these cardinal directions can be used for azimuth; however, traditionally, the horizon closest to the object being observed is chosen. Altitude (or ALT) is easy, for it is just the measure of the angle above or below the horizon specified by the azimuth. Since the horizon closest to the object is always chosen, altitude never exceeds ±90°. From San Francisco (37°47'N/122°25'W), the North Star, Polaris, is located at about +38°alt/0°az or 38° above the northern horizon. On the vernal equinox, one hour <u>before</u> sunrise, the Sun is at a negative altitude, -15°alt/90°az.

The ALT/AZ for an object can be determined if it is on a star wheel. Rotate the star wheel to the desired date and time. Locate the desired object in the window. The periphery of the window represents the horizon, and the center of the window marks the zenith. Draw an imaginary line from the zenith through the object to the nearest horizon. This indicates the direction in which an observer must look to see the object, or its azimuth. Interpolate the actual azimuth value between the two nearest cardinal directions. Most star wheels come with a scale rule for measuring the angular distances between objects represented on the wheel. If a scale rule is not provided with the star wheel, one can be constructed by measuring the shortest distance between Polaris (in the middle of the central grommet) and the northern horizon. If the star wheel is made for 40°N latitude, then the measured distance is equivalent to 40° on the scale. Place the edge of this scale on the imaginary line, drawn between the object and the nearest horizon, so that the 0° mark falls on the periphery of the window and the +90° mark falls on the center of the window. The mark on the scale next to the object indicates the object's altitude. Star wheels are made for many latitudes, so it is possible that an observer's star wheel does not accurately represent star locations in the observer's sky. A star wheel made for 40° N latitude will show all the stars 10° south of their positions as seen from 30° N latitude. For example, at 2200 EST on 20 January, Betelgeuse (Alpha Orionis) appears at +52°alt/180°az for an observer located at 40°N latitude. However, it appears at +62°alt/180°az as viewed from a location at 30°N latitude.

Right Ascension/Declination. On the celestial sphere, RA and DEC are used. If it were not for Earth's motion, RA and DEC would be the same as longitude and latitude, respectively. But Earth rotates (about itself) and revolves (about the Sun), so distant sky objects observed from Earth move across the sky in cycles that have both daily and annual periods. DEC is measured in degrees north or south of the celestial equator. This helps make DEC easy to understand, it is just like latitude, except '+' indicates north and '-' indicates south. Polaris has a DEC value of about +89°, which puts it very near the north celestial pole (+90°). DEC is measured on the star wheel just like altitude, except the 0° mark falls on the celestial equator and the +90° mark falls on the north celestial pole. Also, since DEC is measured independent of the horizons, the DEC values of sky objects do not change for different observer locations or times.

Right Ascension (RA) can seem fairly complicated, but it is easy to learn using a star wheel. The reference direction for RA is the apparent location of the Sun against the background of stars on the date of the vernal equinox. This can be shown on the star wheel as a line from the north celestial pole (NCP) through the ecliptic to the date mark of the vernal equinox (~21 March) on the rim of the wheel. This line is called the Sun-Vernal Equinox line (SVE). The Sun's apparent path through the sky is called the ecliptic, and is labeled on most star wheels. Notice that on the date of the vernal equinox, the ecliptic and celestial equator cross; the Sun is, thus, on the celestial equator on that date. If one rotates the wheel half a year around, one sees that the ecliptic crosses the celestial equator again around 21 September; this is the autumnal equinox. RA is measured in hours eastward (the direction Earth rotates) from the SVE; on the star wheel, this is clockwise. Recall that longitude is measured in degrees (east or west). Longitude could be measured in hours like RA, using the rotational rate of Earth to make the conversion. Since Earth rotates about 360° in 24 hours, 15° is equivalent to 1 hour of rotation, or 1° equals 4 minutes of rotation. It will be useful to remember these numbers. Using these ideas, Washington D.C. would be located at about $(360° - 77°)/15°/h = 18^h52^m$ or 18:52

RA can be determined by setting the star wheel so that the desired object is in the display window and 21 March is aligned with an hour mark (it does not matter which one). Draw an imaginary line from the north celestial pole through the object to the hour marks on the jacket of the star wheel. Count the number of hours clockwise from the 21 March mark to the imaginary line, and interpolate the number of minutes.

The reason RA may seem complicated is that in order to translate the RA/DEC coordinates into ALT/AZ coordinates, one must know the RA/DEC coordinates, the date, the time of day, and the latitude/longitude coordinates of the observer. The SVE line does not move with Earth, but is fixed in space. Since Earth revolves 360° in about 365.25 days, the SVE line appears to move relative to a line from Earth's current position to the Sun just a little less than 1° every day. Drawing the SVE line on the star wheel will help make it seem less complicated. Table 1 provides some examples of RA/DEC coordinates.

Table 1: epoch 2000.0 RA/DEC coordinates of some objects

Name of Object	Right Ascension	Declination
Sun on vernal equinox	00:00:00	0°00'00"
Sun on summer solstice	06:00:00	+23°27'00"
Sun on autumnal equinox	12:00:00	0°00'00"
Antares (Alpha Scorpii)	16:29:24	-26°25'54"
Polaris (Alpha Ursae Minoris)	01:31:48	+89°15'49"
Pollux (Beta Geminorum)	07:45:18	+28°01'33"
Rigel (Beta Orionis)	05:14:32	-08°12'05"
Vega (Alpha Lyrae)	18:36:56	+38°47'00"

Procedures

Apparatus

globe, world maps, and local maps that show latitude and longitude.

A. LAT/LONG Coordinates

1. Use maps of the local area to determine the latitude and longitude of the Astronomy Lab (to the minute-of-arc, if possible). In which time zone is the Astronomy Lab located?

2. What is the longitude time correction for the Astronomy Lab? Hint: use the number of degrees from the longitude at the reference of the time zone.

3. Determine the latitude correction for your star wheel. Hint: you must know for what latitude the star wheel was designed.

4. Use a globe or map of Earth to determine the approximate LAT/LONG coordinates of Mexico City, Mexico; Santiago, Chile; Sydney, Australia; Tokyo, Japan; Nairobi, Kenya; and Berlin, Germany.

B. ALT/AZ Coordinates

1. Find the approximate altitude and azimuth coordinates at the Astronomy Lab for the following objects: the Sun on the vernal equinox at two hours before sunset, Antares (Alpha Scorpii) at 2100 on 14 July, and Deneb (Alpha Cygni) at midnight on 30 July.

2. Find the names of the bright stars appearing at these approximate ALT/AZ coordinates from 30°N latitude: +28°/180° (22:00 06 Oct.), +65°/180° (02:00 18 Jun.), +60°/180° (14:00 17 Jun.). Hint: assume no longitude time correction was made.

3. Determine the ALT/AZ of the Sun at solar noon on the winter solstice (~21 December) as seen from the Astronomy Lab. Hint: the star wheel uses approximate solar time, so solar noon appears as noon on the star wheel.

C. RA/DEC Coordinates

1. Find the approximate right ascension and declination for the following objects: Sirius (Alpha Canis Majoris), Betelgeuse (Alpha Orionis), and Regulus (Alpha Leonis).

2. Find the name and Bayer constellation designation of the bright stars located at the following RA/DEC coordinates: 04:35:55/+16°30'33", 13:25:12/-11°09'41", 05:16:41/+45°59'53".

3. If Sirius (Alpha Canis Majoris) transits the meridian at midnight, when does Pollux (Beta Geminorum) transit the meridian? Hint: determine the difference in RA between Pollux and Sirius. When does Regulus (Alpha Leonis) transit the meridian during that same night?

4. Determine the RA/DEC of the Sun at solar noon on the winter solstice.

Astronomy Laboratory Exercise 3
Sky Patterns

On a clear, dark night, one may look into the sky and wonder at the majesty of the heavens. At first, the stars appear sprinkled randomly across the sky. But, after a little while, the stars seem to be arranged in patterns; stars may make a line here, a circle there, another shape elsewhere. It is only natural that the human mind, used to order and familiar with earthly things, should begin to imagine patterns in the stars. Our ancestors, who spent more time outside under the heavens than we do today, enshrined their stories among the stars by naming sky patterns for heroes, monsters, and other concerns of their time.

Many different cultures have their own collection of sky patterns and associated stories. The pattern of stars we call the Big Dipper has been seen as a wagon, a plow, a casserole pan, the thigh of a bull, a complex procession of celestial bureaucrats (composed of hippopotami, crocodiles, and elephants), and, of course, the tail end of a bear. Today, we see the Big Dipper as part of the larger pattern of Ursa Major (or Greater Bear).

In 1930, the International Astronomical Union divided the entire sky into eighty-eight constellations based on the sky patterns of the ancient Greeks and Romans. Other patterns, called asterisms, are formed from part of a single constellation or from parts of many constellations. The Big Dipper, an asterism, is part of the constellation of Ursa Major. The Summer Triangle, another asterism, is formed by the brightest star of each of three constellations: Lyra, Cygnus, and Aquila.

On clear nights, away from the bright lights of the cities, one can look up at the heavens and see a milky band of light stretching across the sky. This is the Milky Way, a band of stars so plentiful that individual stars often cannot be resolved with the naked eye. This band is a part of the Milky Way galaxy, a spiral collection of over 100 billion stars, of which the Sun is one.

There are many things to see in the heavens. Besides the imaginary sky patterns, one can see comets, planets, individual and multiple star systems, supernovae (exploding stars), clusters of stars, nebulae, other galaxies, and more. The natures of some of these objects have only recently been revealed to us. For example, even in the first part of the twentieth century humans thought of spiral galaxies as nothing other than spiral shaped nebulae (clouds of dust and gas) within our own galaxy. We now know that the Milky Way galaxy is only one of many billions of other galaxies. There are still many mysteries to be solved.

Light and air pollution limit many modern Earth-bound observers from enjoying the night sky. It is often necessary to escape from the city and go out into the country in order to see anything but the brightest objects in the sky. Such a trek into the wild requires some preparation. One should not depend on the weather forecast. A warm day (or even a forecast for a warm evening) does not indicate how the individual will feel. Wear layered clothing to allow for any temperature variation. When the temperature drops outside, the body must use more stored energy to keep the body at a constant temperature; so, bring along a snack. Remember that one is leaving "civilization," so do not expect any facilities. Beverages containing diuretics (such as caffeine and ethanol) are NOT recommended. Insect repellent of the lotion-type may be useful and is preferred to the spray-type, which may get onto telescope lenses, mirrors, and other optical instruments.

Procedures

Apparatus

telescopes and binoculars.

This observing lab is divided into three sections, a Pre-Observing part, an Observing part, and a Post-Observing part. The Pre-Observing part should be done well before the planned observing session so you can establish a game-plan for the actual observing lab. Since some objects may be visible just after sunset, the lab may begin early. Check with your instructor for the appropriate meeting time. Also included is a section for practice with the almanac and the star wheel.

As part of the lab assignment, the student should assist in at least two of the following tasks: loading the vehicles or carts with equipment at the Astronomy Lab, unloading and assembly at the observing site, disassembly and reloading at the site, and unloading at the Astronomy Lab. Note on your lab report the activities that you performed.

A. Pre-Observing

1. Use the almanac to determine the times of sunset and end of twilight (or beginning of twilight and sunrise for morning observing sessions) for the date of the observing lab. Star-gazing cannot begin while the Sun is above the horizon. Twilight is the time just after sunset or just before sunrise in which the sky is too bright to see many stars. Rise, set, and transit times depend on the rotation of Earth, so add a correction for the longitude of the observing site to these times. Since these times also depend on the Sun-Earth geometry, the Equation of Time correction should be included if accurate times are needed.

2. Determine the phase of the Moon and its rise and set times. The Moon can make the sky so bright that many sky objects become invisible in the glare from the Moon. It is best if the Moon is not in the sky during observing times.

3. List a few of the brighter constellations that should be visible at the observing time.

4. List every planet that should be in the sky at the observing time. Determine the constellation in which each planet should reside and the approximate ALT/AZ coordinates at the expected observing time.

B. Observing

Enter the observing site with care, for other observers may have arrived before you. If possible, park your vehicle away from the actual observing site. As you approach the site, turn off your headlights but leave your parking lights on until you park. Either turn off your vehicle's cabin lights or cover them with opaque paper or plastic so they do not disturb other observers. Use only dim, red safe-lights at the observing site.

1. Sign-in at the observing site, and help set up the equipment.

2. Note your arrival time and the time you begin to observe in your report. Is this before or after sunset? Is this before or after the end of evening twilight? Also note whether the Moon is visible as well as the constellation in which it currently resides.

3. Find Polaris in Ursa Minor and note which way is north. Use this starting point to find the other constellations you listed in the Pre-Observing section. Which of these constellations can you find? Sketch two of these constellations. Determine the ALT/AZ of the brightest star in each of these two constellations using hand-angle measurements. Be sure to record the date and time of these measurements.

4. If any planets are visible, note their locations against the background of stars (i.e., identify the constellation in which each currently resides). Note the time and determine the ALT/AZ of these planets using hand-angle measurements. Use optical aids to view the planets under higher magnification. Sketch one planet in detail. Your sketch should include the planet's name, its ALT/AZ, the date and time, the type of telescope, and the magnifications used.

5. Note the locations (give constellation, date, time, and ALT/AZ) of any visible clusters, nebulae, galaxies, et cetera (with or without telescopic assistance). Pick one of these objects to describe (this may include a sketch). Note the date, the time, its ALT/AZ, and the magnifications and telescopes used.

6. Describe any meteors, satellites, or other spacecraft seen during the observing session. You may even want to describe some aircraft that pass by during the observing session.

7. Help pick-up the observing area. Sign-out.

C. Post-Observing

1. Using the star wheel set to the time at which your measurements were made, find the ALT/AZ of the brightest star of each of the two constellations sketched during the observing session. Compare the observed coordinates with the ones from the star wheel (i.e., by computing the percentage error).

2. Compare the observed ALT/AZ coordinates of the planets with the ones predicted in the Pre-Observing section (i.e., by computing the percentage error).

3. List some new constellations and all planets visible one month from this observing session.

Dark Sky Observing

On a clear, moonless night, far from city lights, the stars shine like thousands of crystals in a black velvet sky. Such conditions are called **dark sky**. As the eye adapts to the darkness, other objects, some dim and fuzzy, appear. These are the **dark sky objects** or **DSO's**.

Dark sky objects include star clusters, nebulae, galaxies, quasars, and many other objects that are best or only seen in dark skies. Many of these objects look like nothing more than fuzzy patches even in the largest telescopes. Some can only be seen by using long exposure photography. Yet there are others that appear enormous and bright in a dark sky, and one wonders why they were hidden by the city's lights.

DSO hunting can be challenging and fun. There are even awards for successfully finding and, usually, sketching or photographing the dimmer DSO's. DSO's have caused their fair share of problems for comet hunters. Since many DSO's appear like dim, fuzzy comets in the telescope, comet hunters can lose valuable time by mistaking a DSO for a comet. Comets orbit the Sun, so they can be seen to move against the background of stars from night to night. DSO's, however, are so far away that any proper motion they have may go unnoticed for centuries. Many comet hunters carry lists of DSO's, so as not to mistake one for a comet.

A comet hunter by the name of Charles Messier was one of the first to develop a list of DSO's. He compiled his list of objects looking like comets in the late 1800's, so he would not mistake a DSO for a comet. Messier compiled a list of 103 DSO's, which was later extended to 109. Messier's catalogue can be found in Exercise 5, The Messier List.

A more extensive catalogue, the *General Catalogue of Nebulae*, was published in 1864 by Sir William Herschel and his son, John. This catalogue was unusual in that it included DSO's from both the northern and southern hemispheres. In 1888, J. L. E. Dreyer published an extension of the Herschel catalogue, called the *New General Catalogue of Nebulae and Clusters of Stars* or, simply, the *NGC*. The *NGC* and its supplement, the *Index Catalogues* (or *IC*), contain nearly 15,000 DSO's. The Great Orion Nebula, for example, is M 42 in Messier's list and NGC 1976 in the *New General Catalogue*.

This observing lab is divided into a pre-observing section and an observing section. A successful and rewarding observing session can only be assured by completing the pre-observing part before the planned observing session. To be unprepared for an observing session invites frustration and disappointment.

Procedures

Apparatus

   star charts, various telescopes, eyepieces, binoculars, and red safe-lights.

A. Pre-Observing

All questions apply to the date and location of the observing session.

1. Obtain from the instructor the latitude and longitude of the observing site. Also, obtain the time of the observing session.

2. Determine the times of sunset and the end of twilight for the observing site.

3. Determine the phase of the Moon and its rise and set times for the observing site.

4. List the planets that are visible during the observing time and the constellations in which they reside.

5. Use Table 1 or the Messier list (in Exercise 5) to make a list of the DSO's that will be visible during the observing session. Locate these DSO's on a star chart. This will help you find the DSO's during the observing exercise.

B. Observing

With your star charts and your answers to the Pre-Observing questions, make your way to the observing site. Do not use white light sources at the observing site, as white light destroys night-vision. Automobile headlights and cabin lights should be dimmed before entering the site. Dim, red safe-lights will be available at the observing site.

1. Sign in at the observing site. Describe in your lab report what you did to help with loading the vehicles, setting up the equipment, unloading, etc.

2. Once your eyes have become adapted to the dark, you may notice the glow of city lights or other sources of light pollution. Locate and describe any light pollution that you see.

3. Estimate the ALT/AZ of any visible planets. Estimate their magnitudes by comparing them with nearby stars.

4. Observe and describe the Milky Way with the unaided eyes, then again with a pair of binoculars or a small telescope.

5. Locate the DSO's you listed in part A.5. Using a telescope, sketch these objects *in detail*. Note on each sketch the name, catalogue number, and magnitude of the object. Also, record the telescopes and magnifications you used.

6. If possible, record the image of at least one DSO using film or electronic imaging.

7. Describe any meteors that you see during the lab. Include brightness, color, direction of travel, and constellation of apparent origin (if possible).

8. In your report, if required, describe your overall impression of observing DSO's.

Table 1: Selected Dark Sky Objects which may be visible during the observing session.

Name/Description	Messier	NGC	Constell.	Magnit.	RA/DEC (epoch 2000.0)
Butterfly Cluster	M6	6405	Sco	4.6.	17:40.0/-32°12'
open cluster	M7	6475	Sco	3.3.	17:54.0/-34°49'
Lagoon Nebula	M8	6523	Sgr	5.0.	18:03.7/-24°23'
Wild Duck Cluster	M11	6705	Sct	5.8.	18:50.7/-06°17'
Globular Cluster in Her	M13	6205	Her	5.9.	16:41.5/+36°29'
Eagle Nebula	M16	6611	Ser	6.6.	18:18.9/-13°47'
Trifid Nebula	M20	6514	Sgr	9.0.	18:02.4/-23°02'
Andromeda Galaxy (spiral galaxy)	M31	224	And	3.5.	00:42.3/+41°14'
spiral galaxy	M33	598	Tri	5.7.	01:33.5/+30°37'
open cluster	M34	1039	Per	5.2.	02:41.5/+42°45'
open cluster	M35	2168	Gem	5.3.	06:08.8/+24°20'
open cluster	M36	1960	Aur	6.0.	05:35.6/+34°08'
open cluster	M37	2099	Aur	5.6.	05:51.9/+32°33'
open cluster	M39	7092	Cyg	4.6.	21:31.9/+48°24'
open cluster	M41	2287	CMa	5.0.	06:47.0/-20°46'
Orion Nebula	M42	1976	Ori	3.0.	05:35.3/-05°23'
Beehive Cluster or Praesepe (open cluster)	M44	2632	Cnc	3.9.	08:40.4/+19°41'
Pleiades (open cluster)	M45	--	Tau	1.4.	03:46.5/+24°21'
Sombrero Galaxy (spiral galaxy)	M104	4594	Vir	8.1.	12:40.0/-11°37'
Double Cluster in Per (open clusters)	--	869&884	Per	4.0&4.0.	02:18.5/+57°07' & 02:21.9/+57°05'
planetary nebula	--	7293	Aqr	6.3.	22:29.2/-20°50'

Astronomy Lab Exercise 5
The Messier List

Charles Messier (1730 - 1817) was a French astronomer who specialized in tracking comets. In a telescope, comets too far from the Sun to form pronounced tails appear like fuzzy stars that are little different from unresolved nebulae, galaxies, and star clusters. Comets orbit the Sun, so, unlike objects outside the Solar System, they appear to move against the background of stars. However, since these comets are so far from the Sun, their apparent motions are very small, and it may take several days or weeks to produce any noticeable changes in position. As an aid to himself and other comet hunters, Messier compiled a list of 103 fuzzy objects that did not move against the background of stars. The nature of most of these objects was unknown until the 20ᵗʰ century. In this exercise, students plot the locations of the Messier objects on a star wheel (Figure 5-1, also see Exercise 1, Star Wheels). Armed with a modern understanding of what the Messier objects are, it will be seen that plotting them on a map of the sky reveals much about the nature of the Milky Way galaxy.

Messier was a comet hunter, and did not originally intend to compile a list of **deep sky objects**, such as star clusters, nebulae, and galaxies. Messier discovered 21 comets during his lifetime, but is now better known for his list, which has been so useful to beginning observers that many of these objects are referred to by their numbers in this list. For example, M 31 denotes the Andromeda galaxy and M 42 denotes the Great Nebula in Orion. Objects M 104 through M 109 were added by Messier's colleague, Pierre Méchain. Three of the 109 objects (M 40, M 91, and M 102) were mistakes by Messier. M 40 is only a double star, M91 has never been explained, and M 102 is the same as M 101. Also, M 73 is a multiple star system consisting of only four stars.

William Herschel (1732 - 1822) and his son, John Herschel (1792 - 1871), published a more extensive list of deep sky objects in 1864 called the *General Catalogue of Nebulae*. This catalogue was unusual in that it included deep sky objects from both the northern and southern hemispheres. In 1888, J. L. E. Dreyer published an extension of the Herschel catalogue, called the *New General Catalogue of Nebulae and Clusters of Stars* or, simply, the *NGC*. The *NGC* and its supplement, the *Index Catalogues* (or *IC*), contain nearly 15,000 deep sky objects. The Great Orion Nebula, for example, appears in both Messier's list (M 42) and the *New General Catalogue* (NGC 1976).

The Messier list included in this lab, lists for each Messier object, its Messier number, its NGC number (if it has one), its right ascension and declination coordinates, the constellation in which it may be found, its angular size in seconds-of-arc, and the type of object it is. The following types are provided: open star cluster (OC), globular star cluster (GC), diffuse nebula (DN), planetary nebula (PN), elliptical galaxy (EG), sprial galaxy (SG), and irregular galaxy (IG).

Open star clusters are groupings of young Population I (galaxy disk) stars. These stars are typically young, blue-white stars and may still be surrounded by some of the nebulosity from which they condensed. To the unaided eye, open star clusters may appear nebulous, but a telescope allows an observer to resolve individual stars. Rich open star clusters typically contain over a hundred stars. Clusters with fewer than 50 stars are considered poor. Open clusters spread out over time, so it may be difficult to distinguish an old open star cluster from the general background of stars. The Pleiades (M 45) is a typical example of an open star cluster.

Globular star clusters are old groups of Population II (galaxy halo) stars. There may be hundreds of thousands of stars in a globular cluster, providing enough gravitational attraction that these clusters have survived since the formation of the Milky Way. Through a small telescope, globular clusters appear as small, fuzzy patches of light. Larger telescopes reveal that the cluster stars are more tightly packed in the center than in the periphery. The Hercules Globular Cluster (M 13) is a typical example of a globular cluster.

Diffuse nebulae are found mostly in the disk of the Milky Way and consist of gas and dust. A diffuse nebula is called an emission nebula if it is hot enough to produce most of its own light. These are mostly found around hot, young stars. A diffuse nebula is called a reflection nebula if most of its light comes from reflected star light. Supernova remnants are the hot gas and dust blown out of an exploding star (see Exercise 33, Elements and Supernovae). The Great Nebula in Orion (M 42) is a classic example of an emission nebula.

Planetary nebulae are found mostly in the disk of the Milky Way. They are formed when the outermost layers of a central star are blown into space. Through small telescopes, these nebulae appear as small, round disks, making them look like planets. Larger telescopes often reveal a ring- or disk-like structure in them. The Ring Nebula in Lyra (M 57) is a favorite subject for observers with moderate-sized telescopes.

Elliptical galaxies, **spiral galaxies**, and **irregular galaxies**, are all discussed in IL-CG, Classifying Galaxies. Typically, these galaxies are so far away that they appear as only dim, fuzzy patches of light, even in moderate-sized telescopes. Since these objects are outside the Milky Way, they are only seen above and below the disk of the Galaxy, where dust in the Milky Way does not obscure the view. The Andromeda galaxy (M 31) is the closest large, spiral galaxy to the Milky Way.

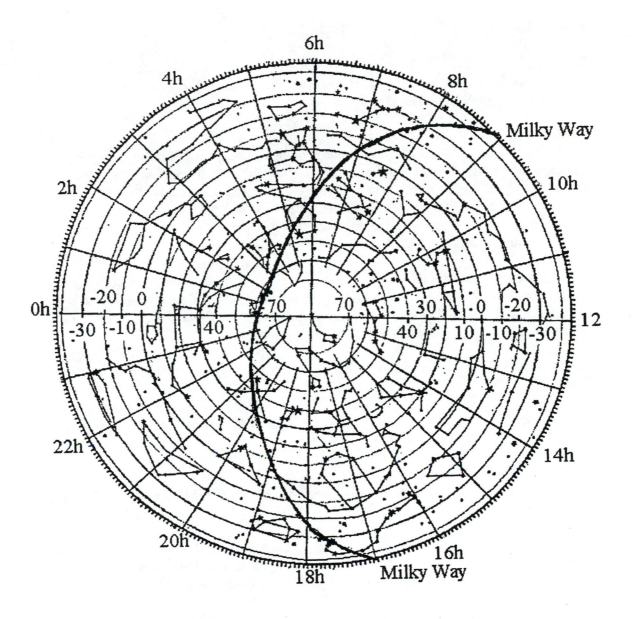

Figure 5-1. A star wheel showing the constellation outlines, Milky Way, hours of right ascension (0 h to 23 h), and declination (-40° to 90°).

Procedures

Apparatus

pencils, rulers, markers of 4 different colors (red, yellow, green, and blue) and photocopy of Figure 5-1, if possible.

A. Plotting the Messier Objects

1. Divide the type of Messier objects between groups of students. Perhaps, one group would have all of the open clusters, another group would have the globular star clusters, and so on. Some care should be taken in doing the division, since there are not equal numbers of objects of each type. Each group should get a marker whose color depends on the type of object being plotted. The suggested color coding scheme is: yellow for open star clusters, red for globular star clusters, green for nebulae, and blue for galaxies.

2. Each group should plot its collection of Messier objects on the star wheel in Figure 5-1 by the right ascension/declination (RA/DEC) coordinates of the objects. Check your work by comparing the constellation in which you plot the object to the constellation given in the Messier list.

3. Finally, use the colored marker representing your object type to highlight the locations of these objects.

4. Share your group's results with the rest of the class in an organized manner.

B. Recognizing Patterns

1. Which types of objects are mostly concentrated in the disk of the Milky Way? Explain why this might be so.

2. Are most of the globular clusters in the disk of the Milky Way? The globular clusters orbit the center of mass of the Milky Way. Based on the distribution of globular clusters, in the direction of which constellation do you expect to be the center of the Milky Way?

3. Which types of objects are mostly concentrated out of the disk of the Milky Way? Explain why this might be so.

Table 1: The Messier List

M	NGC	R.A. h m	Dec. ° '	Const.	Size "	Mag.	Type
1	1952	05 31.5	+21 59	Tau	6 x 4	8.4	DN
2	7089	21 30.9	-01 03	Aqr	12	6.4	GC
3	5272	13 39.9	+28 38	CVn	19	6.3	GC
4	6121	16 20.6	-26 24	Sco	23	6.5	GC
5	5904	15 16.0	+01 16	Ser	20	6.1	GC
6	6405	17 36.7	-32 11	Sco	26	5.3	OC
7	6475	17 50.6	-34 48	Sco	50	4.1	OC
8	6523	18 00.7	-24 23	Sgr	90 x 40	6.0	DN
9	6333	17 16.2	-18 28	Oph	6	7.3	GC
10	6254	16 54.5	-04 02	Oph	12	6.7	GC
11	6705	18 48.4	-06 20	Sct	12	6.3	OC
12	6218	16 44.6	-01 52	Oph	12	6.6	GC
13	6205	16 39.9	+36 33	Her	23	5.9	GC
14	6402	17 35.0	-03 13	Oph	7	7.7	GC
15	7078	21 27.6	+11 57	Peg	12	6.4	GC
16	6611	18 16.0	-13 48	Ser	8	6.4	OC
17	6618	18 17.9	-16 12	Sgr	46 x 37	7.0	DN
18	6613	18 17.0	-17 09	Sgr	7	7.5	OC
19	6273	16 59.5	-26 11	Oph	5	6.6	GC
20	6514	17 59.6	-23 02	Sgr	29 x 27	9.0	DN
21	6531	18 01.6	-22 30	Sgr	12	6.5	OC
22	6656	18 33.3	-23 58	Sgr	17	5.6	GC
23	6494	17 53.9	-19 01	Sgr	27	6.0	OC
24	6603	18 14.0	-18 30	Sgr	90	11.4	OC
25	IC 4725	18 28.8	-19 17	Sgr	35	6.5	OC
26	6694	18 42.5	-09 27	Sct	9	9.3	OC
27	6853	19 57.5	+22 35	Vul	8 x 4	7.6	PN
28	6626	18 21.5	-24 54	Sgr	15	7.6	GC
29	6913	20 22.1	+38 22	Cyg	7	7.1	OC
30	7099	21 37.5	-23 25	Cap	9	8.4	GC
31	224	00 40.0	+41 00	And	160 x 40	4.8	SG
32	221	00 40.0	+40 36	And	3 x 2	8.7	EG
33	598	01 31.1	+30 24	Tri	60 x 40	6.7	SG
34	1039	02 38.8	+42 34	Per	30	5.5	OC
35	2168	06 05.8	+24 21	Gem	29	5.3	OC
36	1960	05 32.8	+34 06	Aur	16	6.3	OC
37	2099	05 49.1	+32 32	Aur	24	6.2	OC
38	1912	05 25.3	+35 48	Aur	18	7.4	OC
39	7092	21 30.4	+48 13	Cyg	32	5.2	OC
40*							
41	2287	06 44.9	-20 41	CMa	32	6	OC
42	1976	05 32.9	-05 25	Ori	66 x 60	4.0	DN
43	1982	05 33.1	-05 18	Ori	20 x 15	9.0	DN
44	2632	08 37.2	+20 10	Cnc	90	3.7	OC
45		03 44.5	+23 57	Tau	120	1.6	OC
46	2437	07 39.6	-14 42	Pup	27	6.0	OC
47	2422	07 34.3	-14 22	Pup	25	5.2	OC

48	2548	08 11.3	-05 38	Hya	30	5.5	OC
49	4472	12 27.3	+08 16	Vir	4 x 4	8.5	EG
50	2323	07 00.6	-08 16	Mon	16	6.3	OC
51	5194-5	13 27.8	+47 27	CVn	12 x 6	8.4	SG
52	7654	23 22.0	+61 19	Cas	13	7.3	OC
53	5024	13 10.5	+18 26	Com	14	7.8	GC
54	6715	18 52.0	-30 32	Sgr	6	7.3	GC
55	6809	19 36.9	-31 02	Sgr	15	7.6	GC
56	6779	19 14.6	+30 05	Lyr	5	8.2	GC
57	6720	18 51.8	+32 58	Lyr	1 x 1	9.0	PN
58	4579	12 35.1	+12 05	Vir	4 x 3	8.2	SG
59	4621	12 39.5	+11 55	Vir	3 x 2	9.3	EG
60	4649	12 41.1	+11 49	Vir	4 x 3	9.0	EG
61	4303	12 19.4	+04 45	Vir	6 x 6	9.6	SG
62	6266	16 58.1	-30 03	Oph	6	6.6	GC
63	5055	13 13.5	+42 17	CVn	8 x 3	10.1	SG
64	4826	12 54.3	+21 57	Com	8 x 4	6.6	SG
65	3623	11 16.3	+13 23	Leo	8 x 2	9.4	SG
66	3627	11 17.6	+13 17	Leo	8 x 2	9.0	SG
67	2682	08 48.5	+12 00	Cnc	18	6.1	OC
68	4590	12 36.8	-26 29	Hya	9	8.2	GC
69	6637	18 28.1	-32 23	Sgr	4	8.9	GC
70	6681	18 40.0	-32 21	Sgr	4	9.6	GC
71	6838	19 51.5	+18 39	Sge	6	9.0	GC
72	6981	20 50.7	-12 44	Aqr	5	9.8	GC
73	6994	20 56.2	-12 50	Aqr	3	9.0	OC
74	628	01 34.0	+15 32	Psc	8 x 8	10.2	SG
75	6864	20 03.2	-22 04	Sgr	5	8.0	GC
76	650	01 39.1	+51 19	Per	2 x 1	11.4	PN
77	1068	02 40.1	-00 14	Cet	2 x 2	8.9	SG
78	2068	05 44.2	+00 02	Ori	8 x 6	8.3	DN
79	1904	05 22.2	-24 34	Lep	8	7.5	GC
80	6093	16 14.1	-22 52	Sco	5	7.5	GC
81	3031	09 51.5	+69 18	UMa	16 x 10	7.9	SG
82	3034	09 51.9	+69 56	UMa	7 x 2	8.4	IG
83	5236	13 34.3	-29 37	Hya	10 8	10.1	SG
84	4374	12 12.6	+13 10	Vir	3 x 3	9.4	EG
85	4382	12 22.8	+18 28	Com	4 x 2	9.3	EG
86	4406	12 23.7	+13 13	Vir	4 x 3	9.2	EG
87	4486	12 28.3	+12 40	Vir	3 x 3	8.7	EG
88	4501	12 29.5	+14 42	Com	6 x 3	10.2	SG
89	4552	12 33.1	+12 50	Vir	2 x 2	9.5	EG
90	4569	12 34.3	+13 26	Vir	6 x 3	9.6	SG
91*							
92	6341	17 15.6	+43 12	Her	12	6.4	GC
93	2447	07 42.5	-23 45	Pup	18	6.0	OC
94	4736	12 48.6	+41 23	CVn	5 x 4	8.3	SG
95	3351	10 41.3	+11 58	Leo	3 x 3	9.8	SG
96	3368	10 44.2	+12 05	Leo	7 x 4	9.3	SG
97	3587	11 11.9	+55 18	UMa	3 x 3	12.0	PN
98	4192	12 11.3	+15 11	Com	8 x 2	10.2	SG

99	4254	12 16.3	+14 42	Com	4 x 4	9.9	SG
100	4321	12 20.4	+16 06	Com	5 x 5	10.6	SG
101	5457	14 01.4	+54 35	UMa	22 x 22	9.6	SG
102*							
103	581	01 29.9	+60 26	Cas	6	7.4	OC
104	4594	12 37.3	-11 21	Vir	7 x 2	8.3	SG
105	3379	10 45.2	+12 51	Leo	2 x 2	9.7	EG
106	4258	12 16.5	+47 35	CVn	20 x 6	8.4	SG
107	6171	16 29.7	-12 57	Oph	8	9.2	GC
108	3556	11 08.7	+55 57	UMa	9 x 2	10.0	SG
109	3992	11 55.0	+55 39	UMa	8 x 5	9.8	SG

* indicates object does not exist

Astronomy Laboratory Exercise 6
About Your Eyes

The human eye is the most widely used astronomical instrument; it is, therefore, an appropriate subject for study in an astronomy lab. Many other astronomical instruments require an intelligent eye. Understanding the functions and limitations of the eye will be helpful in using other astronomical instruments.

Vision involves three distinct steps: (1) a geometrical optics step, in which light from an object is focused on the retina; (2) a detection step, in which rod and cone cells in the retina detect incident light; and (3) a data processing step, in which electrical signals produced by stimulated detector cells are interpreted. The latter step is accomplished by nervous networks in both the retina and brain, and also involves memory functions. This lab will explore the detection step.

Figure 6-1 is an anatomical drawing of the major parts of a human eye ball. The eye focuses light to form images on the retina by refracting (or bending) the incident light with the cornea and the lens. The principal refracting surface of the human eye is the cornea. That is to say, the cornea functions as a lens to focus light on the retina. The part of the eye called the lens serves mainly to adjust the focal length of the cornea-lens system. The lens is said to provide accommodation for distance, meaning that it changes its shape to allow images to be brought into focus on the retina for objects at different distances from the eye. The shape of the lens is controlled by the ciliary muscles in the eye. The shape of the cornea is normally fixed, but can be modified with a contact lens or through surgery. The operation of lenses will be explored in subsequent labs.

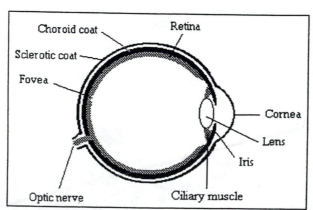

Figure 6-1: Cross section of the human eye as viewed laterally.

The electromagnetic spectrum (the light spectrum) stretches over a wide range, and includes radio, microwave, infrared, visible light, ultraviolet, x-rays, and gamma-rays. These forms of light vary in wavelength. Radio waves have the longest wavelengths, and gamma-rays have the shortest wavelengths. The amount of energy associated with light is inversely proportional to its wavelength. So, x-rays, which have short wavelengths, are more energetic than microwaves, which have long wavelengths. Each region of the spectrum can be divided into "colors." For example, radio is divided into VLF (very-low-frequency), AM (amplitude modulation), VHF (very high frequency), FM (frequency modulation), and UHF (ultra high frequency). Although the visible portion of the spectrum is a tiny part of the whole electromagnetic spectrum, it is subdivided too. We call these subdivisions colors. In order, from longest wavelengths to shortest, the colors of visible light are red, orange, yellow, green, blue, indigo, and violet.

The human eye is most sensitive to light with a wavelength of about 540 nm (green), but responds to light from about 400 nm (violet) to about 700 nm (red). An exercise in this laboratory allows the student to determine the range of his/her vision. Light detection is accomplished by detector cells located in the retina of the eye. There are two basic types of detector cells: rods (rod-shaped) and cones (cone-shaped). Rod cells come in one variety and provide only gray-scale vision (black and white and shades of gray). These detectors are very sensitive to light and work best in dim lighting. Also, rod cells are slow in delivering their message ("light is detected") to the brain. Running through an obstacle course at night, when rod cells are being used, is dangerous because one may run into an obstacle before the brain receives its images from the eyes.

Cone cells come in three varieties: red, green, and blue. The red cone cells detect red light best, but, in order of sensitivity, detect orange, yellow, green, and blue. The green cone cells detect green light best, but also detect other wavelengths of light with varying sensitivity. Similarly, the blue cone cells detect blue best, but also detect some green and violet. Cone cells are not very sensitive to light, so they do not work well in dim lighting. Cone cells provide fast vision by sending their message ("light is detected") to the brain quickly. Cone cells provide humans with "day-vision," while rod cells provide "night-vision." See Figure 6-2 for a graphical comparison of the sensitivity versus wavelength of the detector cells.

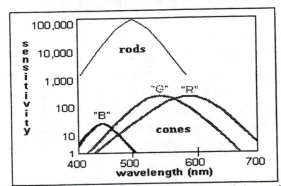

Figure 6-2: The relative sensitivities of the rods and the three sets of cones of the human eye.

Human eyes function in both very bright light and in very dim light. Human eyes perform two adjustments to accommodate to changes in light intensity: changing the pupil size and switching the type of detector cells being used. The size change of the pupil can be seen by placing one's face close to a mirror and making the light alternately dim and bright. In bright lighting, the pupil has a diameter of about 2 mm, but, in dim lighting, the pupil expands to a diameter of about 6 mm. This change in area ($6^2/2^2 = 36/4 = 9$) is roughly ten times. Thus, part of the eyes' increased sensitivity in dim lighting is accomplished by opening the aperture to allow in more light.

A second and much more significant change occurs in the detector cells used. Cone cells are used as light detectors in daylight conditions. These cone cells provide fast, color vision. In dim lighting, however, cone cells effectively stop working for the lack of light. This is why colors seem to wane as twilight descends. If one waits in the dark from 5 to 20 minutes, depending somewhat on the individual, rod cells begin to operate. The graph of Figure 6-2 shows that rod cells are most sensitive to blue light and almost insensitive to red light. Consequently, one finds dim, red "safe-lights" in use at astronomical observing sites. This allows people to see dim sky objects using rod-vision and avoid tripping over equipment and each other using red cone-vision. An exercise in this laboratory allows the student to measure the time required for his/her dark accommodation (or adaptation) to occur.

A single flash of bright light is enough to destroy one's night vision for several minutes. Rod cells contain a photopigment (a molecule that is sensitive to light) called rhodopsin or visual purple. When light encounters a rhodopsin molecule, the molecule photoisomerizes into a form that is insensitive to light. It takes some time for the molecule to return to its light sensitive form. It is a common experience to see an after-image, resulting from, say, a photographic flash. When such a flash occurs, a bright image is formed on the retina, photoisomerizing all or most of the photopigments of the detector cells on which the image falls. It takes some minutes for the detector cells to fully regenerate the photopigments and regain their original sensitivity to light. The after-image is said to be negative if it appears dark where the original image was bright and bright where the original image was dark; colors may also be switched to their complements. A positive after-image occurs if regions of brightness and darkness in the after-image correspond to the same regions in the original image and if the colors are the same.

Eyelids transmit roughly 30 percent of incident light, depending somewhat on blood volume in the eyelid tissue, skin pigments, and the spectral composition of the light. Hemoglobin and myoglobin (red proteins) in the eyelid tissues, cause the eyelids to function as red filters. This means that the eyelids transmit the red end of the visible spectrum best, while most of incident blue light is absorbed. Night-vision can be protected from an expected flash of light by both closing the eyelids and covering the eyes with an opaque object (such as the hands).

The <u>Random House College Dictionary - Revised Edition</u>, 1980, defines "to see" as "1. to perceive with the eyes; look at; 2. to view; visit or attend as a spectator; 3. to perceive mentally; discern; understand ...." Seeing requires a lot more than letting light enter the eyes. Once the light has been detected (by either rod or cone cells), a signal must be sent to the brain for interpretation. Since there are about 130 million detector cells in each eye and only about one million nerve axons leading from each eye to the brain, some data processing must occur in the retina. Once information about an image reaches the brain, a complex series of analyses and associations with memory is used to relate the image to a concept. How we actually see an object depends strongly on the warehouse of memories from which we draw these associations.

Procedures

Apparatus

alpha ray range apparatus, safe-light, and phosphor screen (in box), red, orange, yellow, green and blue color filters, 20 cm x 20 cm, spectrometer, and incandescent bulb with dimmer lamp.

NOTE: Parts A, B and C can be done in any order. Part A should be done by small groups in a room that can be made dark.

A. Observing Alpha Ray Scintillations with Dark Adapted Eyes

Alpha rays are a type of radiation emitted by some radioactive materials that are a natural part of Earth and other bodies in the Universe. Alpha rays have high energy, but travel only a short distance in air, about 3 to 5 centimeters (cm). They are of no health hazard as long as the radiation source is outside your body (so don't eat or inhale alpha emitters!). If a phosphor screen is placed close to a source, the screen can be seen to scintillate (give off a tiny flash) at each point and every time an alpha particle hits it. Dark observing conditions and night vision are required to see these scintillations.

1. The phosphor screens should be in a light-tight box. Do not open the box when the room lights are on, else the phosphor screens will glow too brightly to be used for several hours. Do this exercise in a room that can be made dark. With the door closed, turn on the overhead lights so your eyes are adjusted to bright light. Set up the alpha-ray apparatus in the middle of the room. Also, set up a very dim light source in a corner of the room. Make sure that the alpha source is about 2 cm from where the phosphor screen will be placed. Note the time and turn off the overhead lights.

2. With the bright lights off, remove the phosphor screen from its box and place it just behind the eyepiece, with the phosphor toward the alpha source. Sequentially check for scintillations by looking through the eyepiece for just a moment. If you do not see the scintillations, step away and allow the next student to look. Repeat every minute or so, until scintillations are seen. Record the time when you first see the scintillations. There will be uncertainty about just what is expected until the scintillations are seen; once seen, however, it should be clear. Typically, 5 to 10 minutes are required to see the scintillations. An additional 5 to 10 minutes may be required to achieve full night-vision, at which time the scintillations will cover your entire field of view. Record this time. This is your dark adaptation time. Briefly describe what you see. Where in your field of view do the scintillations first appear? Is there any color involved in the tiny flashes?

3. After all members in the group have seen the scintillations (or given up after a reasonable amount of time), place a piece of paper between the alpha source and the phosphor screen to verify that one sheet of paper can stop alpha rays.

4. Make a table listing your dark adaptation time and the dark adaptation times of at least four other individuals. What is the arithmetic mean of these times?

5. Put the phosphor screens back in their box before turning on the room lights, else the bright light will spoil the screens for additional experiments this day.

B. Seeing Color Through Your Eyelids

1. Seat yourself at a lab table, relax, and stare into a bright lamp. Then, close and cover your eyes with your flat hands. You should see an after-image. You may need to repeat this several times, paying close attention to what you are seeing. Describe the after-image. Is it a positive or a negative image? Is the after-image positive or negative if you leave your eyes closed but remove your hands?

2. With your head positioned to stare at a lamp, and with your eyes closed, pass a flat hand across your face very close to your face to cover and uncover your eyes. Repeat several times. You should be able to see the shadow of your hand through your eyelids. Open your fingers and fan them just over your eyelids. Briefly describe what you see.

3. Shade both closed eyes with your hands for 20 seconds, then move your hands, keeping your eyelids down. What colors do you see? Is the light bright?

4. Have your lab partner randomly choose a color filter and place it over your eyes, ensuring that the filter completely shades your eyes. Can you determine the color of the filter with your eyes closed? Repeat for different color filters. You should be able to tell Red from Green or Blue. With some practice, many students can also tell Red from Orange or Yellow. Briefly describe your results.

5. Are results of this section consistent with the eyelids acting as red filters? Summarize evidence to fortify your answer.

C. Determining the Range of Your Vision

A spectrometer is a device that splits incident light into its constituent colors and allows the user to measure the wavelengths of these colors.

1. Set up the spectrometer so that light from a bright incandescent bulb passes through the device. Place your eye at the eyepiece and view the spectrum of visible light.

2. Measure the wavelengths of the ends of the visible spectrum. In other words, determine the wavelength of the border between red and infrared and the wavelength of the border between violet and ultraviolet.

3. Measure the wavelengths of the centers of the basic colors (red, orange, yellow, green, blue, and violet).

Geometrical Optics

   Much of what we know about the Universe comes to us via the electromagnetic spectrum (light). One of the most common devices for exploring the heavens is the optical telescope. There are many different types of optical telescopes. Some use lenses to collect light (refractors), others use mirrors (reflectors), and still others use combinations of lenses and mirrors.. The study of how light is collected and focused is part of the domain of geometrical optics. This lab will explore the operation of a lens.

   The type of a lens is defined by its shape. The most common type of lens used in telescopes is biconvex (or double convex), where the lens curves outward in the middle on both sides. This type of lens, called a converging lens, tends to bend light so that light from a distant object is focused (converges) at a point on the opposite side of the lens. A biconcave (or double concave) lens, in which both sides curve inward in the middle, is sometimes used in telescopes as well. This type of lens tends to diverge light that passes through it, so it is called a diverging lens.

   Related to the shape of the lens is its focal length. Focal length is an intrinsic property of a lens. The lens formula, Equation 1, shows the relationship between the distance of the object from the center of the lens (o), the distance of the focused image from the center of the lens (i), and the focal length (f). Object distance, image distance, and focal length are measured along an imaginary line that runs through the center of the lens and perpendicular to the face of the lens. This line is called the optical axis. When the object distance varies, the image distance changes according to Equation 1. The focal length is a constant for a given lens and color of light.

$$\frac{1}{f} = \frac{1}{o} + \frac{1}{i}, \tag{1}$$

In this laboratory exercise, a bright object will be arranged so that it is centered on the optical axis of a lens. The object may then be moved along the axis, and will produce images at different distances consistent with the lens formula. The sign of i is positive when the image is on the opposite side of the lens from the object, and it is negative when the image appears on the same side. If the object and image distances are known, the focal length of the lens may be computed.

Example 1: An object is placed 20.0 cm from a lens. On the opposite side of the lens from the object, an image is formed at a distance of 30.0 cm from the lens. What is the focal length of the lens?

   First, we solve Equation 1 for f, then substitute values for the variables:

$$f = \frac{o \cdot i}{o+i} = \frac{(20.0 \cdot cm)(30.0 \cdot cm)}{(20.0 \cdot cm + 30.0 \cdot cm)} = \frac{600 \cdot cm}{50.0 \cdot cm} = 12.0 \cdot cm$$

Thus, this lens has a focal length of 12.0 cm.

Images come in two varieties: real and virtual. A converging lens can produce both types of images. A real image is produced when an object is placed farther from the lens than the lens' focal length. Light passing through the lens is bent so as to come to focus at some distance from the object on the opposite side of the lens. If a screen is placed at this position, a focused, inverted, real image appears on the screen. If the object is moved closer to the lens than the focal length, then the light from the object no longer comes to focus; instead it diverges. The eye normally collects light diverging from objects and forms real images on the retina, which the brain then interprets. A converging lens decreases the divergence of the light from the object. The brain interprets this as light coming from a larger object still on the same side of the lens as the original object. The image seen by the eye is a virtual image, and the image distance (i) is negative. A lens used this way is often referred to as a magnifying glass. Another example is the virtual image seen in a planar (flat) mirror, which has a magnification of one.

Magnification is defined as the ratio of the image size (h_{img}) to the object size (h_{obj}). Equation 2 is a formula for finding the magnification produced by thin lenses.

$$M = \frac{h_{img}}{h_{obj}} = \frac{i}{o}, \tag{2}$$

where i is the image distance and o is the object distance. Notice that a magnification of one means the image and object are the same size, and a fractional magnification describes an image that is smaller than the object. Also, notice that since a virtual image always has a negative image distance, magnification of virtual images is always negative. This is interpreted to mean that the image is upright.

Example 2: A 10. cm high object is placed 15 cm from a lens with a focal length of 10.0 cm. Determine the height of the image.

We must determine where the image will appear by using Equation 1.

$$i = \frac{o \cdot f}{o - f} = \frac{(15 \cdot cm)(10.0 \cdot cm)}{(15 \cdot cm - 10.0 \cdot cm)} = 30. cm$$

Now, using Equation 2, the magnification of the image is,

$$M = \frac{i}{o} = \frac{30. cm}{15 \cdot cm} = 2.0 .$$

Finally, the height of the image is,

$$h_{img} = M \cdot h_{obj} = (2.0)(10. cm) = 20. cm .$$

The lens formula and magnification formula work equally well for thin, diverging lenses. However, diverging lenses have negative focal lengths and, by themselves, always produce virtual images.

Ray-tracing is a graphical method of approximating the position of an image given the position of the object and the focal length of the lens. See Figure 7-1 for examples. Begin by drawing a line segment on a piece of paper to represent the optical axis. At the midpoint of the optical axis, draw a short, perpendicular line segment to mark the middle of the lens (you may draw the lens shape around this line segment if you wish). Place a dot on the optical axis on each side of the lens at a distance in scale with the focal length of the lens. Label these points with an 'f' to indicate that these are the focal points of the lens. Draw an arrow with its tail on the optical axis and its head a perpendicular distance above the optical axis. This arrow represents the object, so it should be placed at a distance proportional to the given object distance. It is necessary to trace the paths of only two rays to determine where the image will fall.

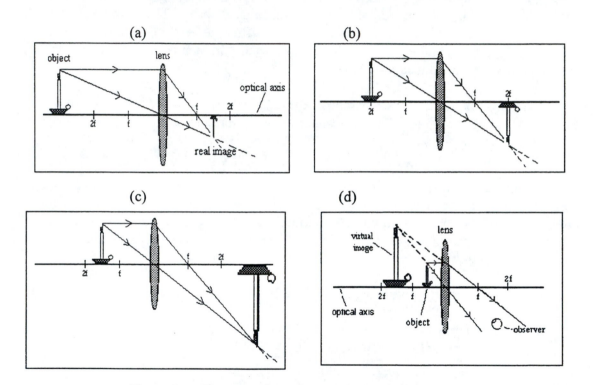

Figure 7-1: The operation of a converging lens.
(a) real image: object distance greater than 2f
(b) real image: object distance equal to 2f
(c) real image: object distance between f and 2f
(d) virtual image: object distance less than f.

The following instructions describe the ray-tracing for producing a real image with a converging lens. Draw the first ray leaving the head of the arrow parallel to the optical axis. It is bent at both surfaces of the lens so as to pass through the focal point on the opposite side of the lens from the object. The second ray leaves the head of the arrow and passes straight through the center of the lens. A third ray may sometimes be drawn. It leaves the head of the arrow, passes through the focal point on the same side as the object and continues until it reaches the lens. It is bent at the lens so as to come out parallel to the optical axis on the other side. The point of intersection of these rays is the location of the image. Notice that the rays do not stop at the point of intersection, but continue on forever unless obstructed.

Light can be focused by two different mechanisms: refraction as through lenses, and reflection as from curved mirrors. Some other important phenomena concerning light are scattering, absorption, and emission. The effects of these phenomena on a beam of light are shown in Figure 7-2. First, the light beam intersects the surface of a transparent block, where it undergoes specular and diffuse reflections. Specular reflections occur with smooth surfaces (like mirrors), and light is reflected at a precise angle. Diffuse reflections occur when light is incident on a rough surface, and light reflects at many different angles. As the light passes through the surface, it bends due to refraction. While passing through the surface, the light is scattered at different angles by minute inhomogeneities in the substance. The substance along the beam is heated as the light is absorbed. Some of the absorbed light may be emitted later. The remaining light reaches the second surface, where, again, it undergoes specular and diffuse reflections. Finally, the beam is refracted out of the block.

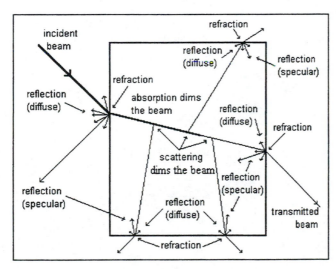

Figure 7-2: The adventures of a beam of light passing through a block of a partly transparent substance.

Understanding these phenomena allows one to explain the appearance of many everyday objects. A single sheet of typing paper, for example, reflects (diffusely) and scatters light so strongly that the transmitted beam is virtually eliminated. A clear window glass produces weak beams by specular reflection, but a strong transmitted beam.

An important characteristic of an optical system, such as a telescope, is its light-gathering power. It is a measure of how much light can be collected by the surface. Light-gathering power is directly proportional to the area of the light-collecting surface. For a circular lens or mirror, the light-collecting surface has an area proportional to the square of the diameter. Equation 3 shows this relationship.

$$LGP \propto A \propto D^2. \tag{3}$$

Example 3: Consider two lenses, one with a 10. cm diameter and the other with a 20. cm diameter. Clearly, the 20. cm diameter lens collects more light than the 10. cm diameter lens. About how many times more light is collected by the 20. cm diameter lens than by the 10. cm diameter lens?

This question can be answered by establishing a set of ratios. Thus,

$$\frac{LGP_{20}}{LGP_{10}} = \frac{A_{20}}{A_{10}} = \left(\frac{\pi r_{20}^2}{\pi r_{10}^2}\right) = \left(\frac{D_{20}}{D_{10}}\right)^2 = \left(\frac{20.\,cm}{10.\,cm}\right)^2 = 4.0.$$

Thus, with twice the diameter, the larger telescope collects four times the amount of light.

Procedures

Apparatus

ruler, optical bench, converging lens (fl. ~15 cm, dia. ~50 mm),
illuminated object, optical bench, clamps, screen, and masks with 5, 10 and 20 mm
diameter holes.

A. Real Images from a Converging Lens

Arrange an optical bench so the long scale can be read conveniently. Place the
illuminated object in a carriage on the optical bench near one end so it can be plugged into
an electrical outlet. Place a screen in a lens clamp and in a carriage at the opposite end of
the bench. Place the provided lens in a clamp between the object and screen, and adjust the
height of the lens, object, and screen so their centers are about the same height above the
bench. Then, position the screen 100.0 cm from the object.

1. Find the two places where the image is in focus on the screen by sliding the lens back and
forth between the object and the screen. Make a table giving the object and image distances
(in centimeters, measured to the millimeter) where focus is achieved. Determine a set of o
(object distance) and i (image distance) values for each of the two places where the image is
in focus.

2. Use Equation 1 and these o and i values to determine f (focal length) for each set. For
each lens, there is only one focal length. If the focal lengths (determined above) are
different, compute the arithmetic mean of the two values of f and use this value for the focal
length in the remainder of the lab.

3. Place the screen a distance of 3f from the source. There should now be no position for
the lens when the image is in focus on the screen. Verify this experimentally.

4. Place the lens half way between the screen and source. Move the screen away from the
source and lens by a couple of centimeters, and again center the lens between the screen and
source. Observe if the image is in focus on the screen. Repeat this step until the the image
is in focus. This procedure should determine the minimum distance between the source and
focused image. Record the o and i distances, and note that they are the same (or nearly so).

5. Notice that the lens formula is satisfied when both o and i equal 2f (i.e., $1/(2f) + 1/(2f) =$
$1/f$). Verify the minimum distance determined in step 4 is equal to 4f. Verify that focus can
not be achieved when the object and screen are closer together than 4f. Comment on your
results.

6. With the source and screen 4f apart, move the lens so that the image is slightly out of
focus. Measure and record the diameter of the lens. Reduce the effective lens diameter by
holding the 10 mm mask directly before the lens, and note all changes observed in the image.
Measure the diameter of the aperture of the mask to the nearest millimeter. Compute the

ratio of light-gathering power for the lens before and after the diameter reduction. Describe all of the effects on the image. Repeat this with all of the masks provided.

7. Notice that the hole in the mask does not need to be in the center of the lens (on the optical axis) for this to work. What is the effect of reducing the effective lens diameter on the range over which an image appears in focus? Is the image brighter with a large diameter or small diameter lens? Is the image easier to obtain focus with a large diameter or small diameter lens?

8. Based on these results, what do you think are the main differences between the views through a large-diameter lens telescope and a small-diameter lens telescope? Which telescope would be preferred for astronomical work? Hint: remember that most of the objects observed in the heavens are very dim as seen from Earth.

B. Virtual Images from a Magnifying Glass

Orient the optical bench so that you can position your eye over the optical bench and look down the optical axis at an object positioned at the other end. Remove the screen and its holder.

1. Position the converging lens (in a clamp) a few centimeters from the object and look into (through) the lens. Observe the magnified illuminated object as you slide the lens away from the object, keeping your eye about 6 cm from the lens. Repeat this several times. Describe the image (if one exists) when $o < f$, $o = f$, and $o > f$. How does this compare with what one expects based on the lens formula? It may helpful to refer to the ray diagrams in Figure 7-1, expecially (d), to think through what is happening here.

2. With your eye 6 cm from the lens, what is the greatest distance between the lens and the object at which the virtual image is still in focus? This distance should be very close to, but less than, the focal length. How does this o value compare to the focal length you found in the previous section (i.e., calculate the percentage error, using the focal length you found in the previous section as the actual value)?

C. Magnification

Arrange the optical bench as in part A. Place the lens at a distance of 40.0 cm from the object.

1. Compute the image distance for the above setup, using $o = 40.0$ cm and the focal length found in part A. Place the screen at this location. Make any minor adjustments to achieve focus, and record the final distance of the screen from the center of the lens.

2. Compute the magnification of this arrangement using the measured o and i values and Equation 2.

3. Measure the size of the image and the size of the object to the nearest millimeter, using a ruler. Calculate the ratio of these two measurements to compute the magnification.

4. Compare these two values for magnification by computing the percentage difference between them.

5. Repeat the above steps using 30.0 cm for o.

Telescopes I

Optical telescopes are the most widely used astronomical instruments. Telescopes provide four advantages for viewing remote objects: they 1) gather more light, 2) magnify the view, 3) provide greater resolution, and 4) provide convenient arrangements for accurately aiming the telescope and measuring angular positions. In optical telescopes, light can be focused by refraction, as through lenses, and by reflection, as from curved mirrors. This lab explores some of the considerations of telescope design and use. More information on telescopes can be found in Exercise 9, Telescopes II.

Probably the easiest telescope to understand is **the astronomical refracting telescope**, illustrated in Figure 8-1. It uses two converging lenses. Light from a distant object passes through the objective lens and forms an intermediate real image inside the telescope. This real image then forms the "object" for the second lens. The second lens, called an eyepiece, acts as a magnifying glass to view this intermediate image. An observer looking into the eyepiece will see an inverted, virtual image that appears much larger than the original object. This telescope is focused by adjusting the distance between the two lenses. If the object is very far away, the intermediate image is formed at about the objective lens' focal length, f_o, from the objective lens. The telescope will be in focus when the eyepiece lens is at about its focal length, f_e, away from that image. Thus, the expected length of the astronomical refracting telescope is, approximately, the sum of the focal lengths, or

$$L = f_o + f_e. \tag{1}$$

Example 1: An astronomical refractor has a 710 mm objective focal length. If a 20. mm focal length eyepiece is used, what is the length of the telescope?
$$L = f_o + f_e = 710 \cdot mm + 20. \cdot mm = 730 \cdot mm.$$

Example 2: Cross hairs are to be added to an astronomical telescope. Where can they be added so they are in focus when observing through the telescope?
> The cross hairs can only be added at the site of the image formed by the objective lens.

Note: For clarity, the angles shown in the following figures are enlarged from what they would be in typical telescopes.

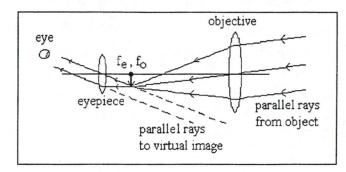

Figure 8-1: The basic components of an astronomical refractor, showing the formation of the intermediate image between the lenses.

The apparent size of an object depends on the size of its image on the eye's retina. That is, the number of detector cells covered by the image is a measure of the object's apparent size. When an object is brought closer, its angular size increases. This increases the size of the image and the number of detector cells covered by its image. A telescope is a device that increases the size of the image of a distant object by spreading the image over more detector cells.

The reason a magnified image is preferable is illustrated by considering the number of pixels that are used when viewing the Moon. A **pixel** is the smallest element of a picture that can be resolved. For the unaided human eye a pixel is about 1 minute-of-arc. Observations of the Moon, which spans about 30 minutes-of-arc as seen from Earth, require an array of about 30 by 30 pixels or 900 pixels. When magnified ten times, 300 by 300 pixels or 90,000 pixels are required. The eye can discern many more details using 90,000 pixels than it can with only 900.

Angular magnification is defined as the ratio of the angular size of the image to the angular size of the object. In Figure 8-2, the angular size of the object, θ_{obj} , is the angle formed by the optical axis and the ray passing through the center of the objective lens to the intermediate image. The angular size of the image, θ_{img}, is the angle formed by the optical axis and a ray passing from that image through the center of the eyepiece lens. Since, in telescopes, these angles are small, the small-angle approximation may be used, i.e.,

$$\tan(\theta) \approx \theta ,$$

where θ is the angle measured in radians. From triangles formed by the optical axis, internal image, and the light rays shown, the angular sizes can be seen to be,

$$\theta_{obj} = \frac{h}{f_o} \quad \text{and} \quad \theta_{img} = \frac{h}{f_e} ,$$

where h is the height of the intermediate image. Magnification is the ratio of the angular size of the image (θ_{img}) to the angular size of the object (θ_{obj}),

$$M = \frac{\theta_{img}}{\theta_{obj}} = \frac{f_o}{f_e} . \tag{2}$$

Thus, the magnification in a telescope is given by the ratio of the objective focal length to the eyepiece focal length. Notice that magnification has no units. Since the eyepiece lens is usually much smaller than the objective lens, it is the eyepiece that is changed for different magnifications. Also, notice that smaller eyepiece focal lengths produce higher magnifications.

Example 3: What is the magnification produced by a 20. mm focal length eyepiece on a telescope with a 1.2 m objective focal length?

$$M = \frac{f_o}{f_e} = \frac{1200 \cdot mm}{20 \cdot mm} = 60.$$

Thus, the image is 60. times larger.

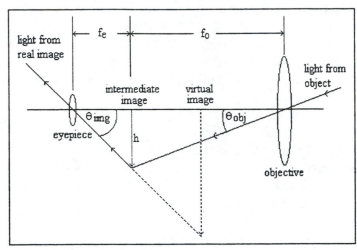

Figure 8-2: Light rays used to show the angular magnification of an astronomical refractor.

Four difficulties occur with a telescope made from two simple converging lenses. First, the focal length for any given lens is slightly different for different wavelengths of light. This problem is called **chromatic aberration**, and results in the appearance of false colors in the image. Chromatic aberration can be nearly corrected by using certain combinations of lenses in place of a single, simple lens. Such a lens system is called an **achromat**. Second, it is difficult to make the objective lens very large in diameter (to collect lots of light and to limit diffraction, which spoils an image) without also making it very thick and heavy. Thick and heavy lenses are difficult to support and expensive to make. Third, a long objective focal length is needed to attain high magnification, but this makes the telescope itself long. A long telescope tube increases the cost of the telescope and makes it difficult to use, move, and house. Fourth, the final image is inverted, which may create confusion when the telescope is used to view objects on Earth.

This last difficulty is resolved by the terrestrial or **Galilean refracting telescope**, illustrated in Figure 8-3. Galileo Galilei constructed this type of telescope in the early 1600's after hearing of a similar design by a Dutch lens maker. The Galilean telescope uses a converging objective lens, but a diverging eyepiece lens. The lens formula, $(1/o) + (1/i) = (1/f)$, also applies to diverging lenses, only f is given a negative value. The use of a diverging lens with a nearby object is shown in Figure 8-4. Such a lens by itself might be referred to as a reducing glass, to contrast its function with that of a magnifying glass. In the Galilean telescope, the eyepiece intercepts light from the objective lens before it can make a real image. This makes the would-be real image of the objective lens a virtual object for the eyepiece. The eyepiece then creates an upright and virtual final image, as shown in Figure 8-3. This may seem confusing at first, but hopefully working through the ray-tracing diagrams with some thought will clarify how this telescope works. The length of this telescope, focused on a distant object, is also given by the sum of the focal lengths, Equation 1, but the eyepiece focal length is negative in this case. The magnification is again the ratio of the focal lengths, Equation 2. Notice that with a negative eyepiece focal length, the magnification becomes negative. A negative magnification indicates that the image is upright.

Example 4: Compute the length and magnification of a Galilean telescope whose objective lens' focal length is 710 mm and whose eyepiece is -20. mm.

$$L = f_o + f_e = 710 \cdot mm + (-20 \cdot mm) = 690 \cdot mm$$

$$M = \frac{f_o}{f_e} = \frac{710 \cdot mm}{-20 \cdot mm} = -36$$

Thus, this telescope is a bit shorter than the astronomical refractor and produces an upright image.

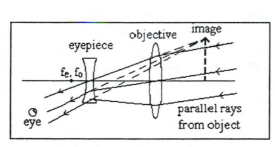

Figure 8-3: The basic components of a Galilean telescope.

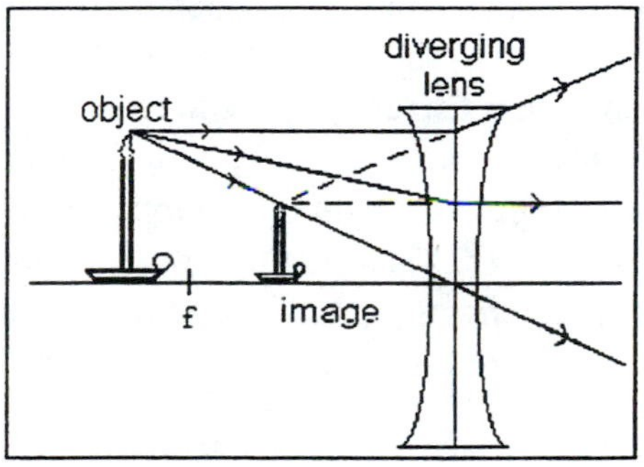

Figure 8-4: A diverging lens.

The inverted image of the astronomical refractor may be made upright by using a combination of internal prisms. Most binoculars use this latter arrangement. This also shortens the total length of the binoculars, but absorbs more of the light, making the final image dimmer than if the prisms had not been used.

Another class of telescopes, the **Newtonian reflecting telescope**, uses a spherically shaped, concave primary mirror to focus the light. It was originally designed and constructed by Sir Isaac Newton in 1668. Newton mistakenly believed that the problem of chromatic aberration could not be solved with glass lenses, so he devised a curved mirror to focus the light. The focal length of a spherical mirror is half of the mirror's radius of curvature. This value is the same for all colors (wavelengths) of light; thus, mirrors do not suffer from chromatic aberration. A planar, secondary mirror is placed in the light path just before the real image comes to focus. It is aligned diagonally to the path of the light, and redirects the image to a hole in the side of the tube. An eyepiece lens is then positioned at the hole to view the image.

Magnification is given by the same formula used with refractors, Equation 2. The length of the telescope tube, however, is about the same as the focal length of the primary mirror. Examples of some types of reflectors are shown in Figure 8-5. In some arrangements the observer is at the upper end of the telescope, which, for large telescopes, adds the inconvenience of requiring a ladder just to reach the eyepiece. A reflecting telescope that employs a spherical mirror suffers from spherical aberration, which prevents the periphery and the center of the image from being in focus at the same time. This is not very apparent in small Newtonians, but spherical aberration can be severe in large telescopes. A parabolic mirror can be used to prevent spherical aberration. But, parabolic mirrors are more difficult to make, and thus, are more expensive.

An alternative design for the Newtonian, developed by L. Cassegrain around 1672, uses a convex secondary mirror to redirect light through a hole in the middle of the primary mirror. An eyepiece may then be placed near the bottom of the telescope. The curved secondary mirror allows the effective focal length of the telescope to be about four times the length of the telescope tube, which makes the telescope less expensive and easier to move. Another change in the Newtonian was made in the 1930's by Bernhard Schmidt. This design uses a refracting corrector plate in front of the spherical primary mirror to correct for spherical aberration. Such telescopes that employ both mirrors (catoptric) and lenses (dioptric) for primary light collection are called **catadioptric**. A **Schmidt-Cassegrain telescope** (or **SCT**) is a popular, moderately priced, catadioptric telescope using both Schmidt's modification and Cassegrain's modification. A diagram of a Schmidt-Cassegrain appears in Figure 8-6.

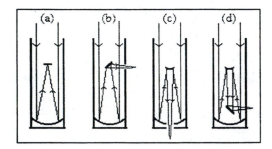

Figure 8-5: Various focus arrangements for reflecting telescopes:
(a) prime focus, (b) Newtonian focus,
(c) Cassegrain focus, (d) Coudé focus

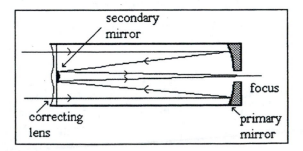

Figure 8-6: Schmidt-Cassegrain telescope.

The **maximum effective magnification** available on a telescope is determined by the diameter of the primary focusing surface. As magnification on a telescope increases, the field of view decreases. Since no more light is collected from the zoomed-in area, an object that is relatively bright at one magnification will appear dimmer at a higher magnification. One can imagine that at some magnification the image will be too dim to easily view. A rule of thumb for the magnification limit of a telescope is given by the 20X per centimeter rule:

$$M_{max} = \frac{20X}{cm} \cdot D, \qquad (3)$$

where D is the diameter of the primary light-collecting surface in centimeters. The 20X is an approximate value, but, for purposes of precision, the 20X may be taken as an exact number. The diameter of the objective lens limits the magnification on refractors. On reflectors, the diameter of the primary mirror determines this.

The 20X per centimeter rule is a good tool with which to check the power claims of telescope advertisements; a telescope with a 60 mm aperture, obviously, can not have a useful power of 500X.

Example 5: A telescope has an objective aperture (which is approximately the same size as the primary focusing surface) of 100. millimeters in diameter. What is the maximum effective magnification of this telescope?

First, the telescope aperture must be converted to centimeters. 100. millimeters is the same as 10.0 centimeters. Using the 20X/cm rule,

$$M_{max} = \frac{20X}{cm} \cdot 10.0 \cdot cm = 200.X.$$

So, 200.X is the maximum effective magnification of this telescope.

A quantity known as the focal ratio (or f-ratio) is used in many fields involving optics, such as photography and astronomy. In photography, the f-ratio is adjusted for different lighting conditions and exposure times. In astronomy, the f-ratio is sometimes used as a rating for telescopes, since telescopes with similar f-ratios produce similar images. The f-ratio is simply the objective focal length divided by the diameter of the objective aperture, or

$$\text{focal ratio} = \frac{f_o}{D}. \qquad (4)$$

Example 6: The f-ratio of a telescope with a 15-cm aperture and a focal length of 1.50 m is

$$\text{focal ratio} = \frac{150 \cdot cm}{15 \cdot cm} = 10.,$$

which is written as f/10.

Telescopes with f-ratios less than f/6 tend to produce bright images and have wide fields of view at low magnifications. These telescopes are well-suited for dark sky objects like nebulae, star clusters, and galaxies, which appear large in our sky but are very dim. Telescopes with f-ratios between f/6 and f/10 provide bright, clear views over a wide range of magnifications. Large f-ratio telescopes are good for lunar and planetary observations, but suffer from low image brightness and small fields of view.

Procedures

Apparatus

colorful objects for viewing through telescopes (like breakfast cereal boxes),
lamp dimmer with clear-glass bulb, filters of various colors,
lenses:　　convex,　　(#1) (100 mm dia., 500 mm fl),
　　　　　　　　　　(#2) (50 mm dia., 150 mm fl),
　　　　concave,　(#3) (50 mm dia., -150 mm fl),
optical bench, 3 carriages, lens holders, screen,
optical mask (~10 mm dia. aperture), meter stick, and wooden blocks.

A. Chromatic Aberration

Place a lamp on the opposite side of the lab at least four meters from where your optical bench is situated. Use wooden blocks so the lamp is at the same height as the lens on your optical bench. Turn on the lamp. Use a lens holder and carriage to place the 100 mm diameter lens (#1) onto the optical bench. Obtain a real image of the lamp filament on the screen. Adjust the brightness of the lamp for best viewing.

1. Move the screen through focus, and note the colored edges which appear on the image before and beyond focus. Record the distance between the lens and screen at the best "white" focus. With a distant object, the image should form at about the same distance as the focal length. Compute the percentage error between this image distance and the published focal length of this lens (obtained from the instructor).

2. Hold a red filter so all light passing through the lens must also pass through the red filter. Bring the image to its best "red focus," and record the "red image distance." Repeat with the blue filter. Where would you expect the system to be in focus with the green filter? Why? Try it and see if it is. Where would you expect the focus to be with infrared light (a color invisible to the human eye but very useful in astronomy)? Briefly explain in your own words what chromatic aberration is, using the results from this experiment to illustrate.

3. With an image of the lamp filament still on the screen, reduce the effective diameter of the lens to 10 mm by using the 10 mm mask. Is the chromatic aberration still visible? Adjust the focus to see if it appears. Also, experiment with the positioning of the 10 mm mask so it is off the optical axis and near the edges of the lens. Describe and briefly explain what is seen.

B. Making Telescopes
1. The Astronomical Refractor

Construct an astronomical refractor by placing the 50 mm diameter, convex eyepiece lens (#2) a distance behind the 100 mm diameter, convex objective lens (#1). Adjust for focus by looking through the eyepiece and objective lenses at a colorful object on the opposite side of the room. Slowly increase the distance between the objective and eyepiece lenses until just before the image loses focus.

a. Is the view magnified? You should be able to observe that there is magnification by looking at the object directly and then through the telescope. You may notice that you can see some details through the telescope which are not visible without it. Describe the view through the telescope. Is the image upright or inverted? Is spherical or chromatic aberration visible? Describe the view through the telescope.

b. What is the distance between the lenses at which the view is most magnified and clear? This is the actual length of the telescope. Compare the expected telescope length to the actual length (i.e., calculate the percentage error, using the measured length as the actual value; you are testing Equation 1).

c. Compute the magnification of the telescope using Equation 2.

2. The Galilean Refractor

Construct a Galilean refractor by using the 100 mm diameter, convex lens (#1) as the objective and the 50 mm diameter, concave lens (#3) as the eyepiece. Slowly increase the distance between the objective and eyepiece lenses until just before the image loses focus.

a. Does the telescope magnify the view? Does it produce an upright image? Is there spherical or chromatic aberration? Describe the view through the telescope.

b. Record the actual length of the telescope (the distance between the lenses) when the image is clear and most magnified. Compare the expected telescope length (Equation 1) to the actual value. Remember that the focal length of a diverging lens is negative.

c. Calculate the magnification produced by this telescope.

C. Exercises with Telescope Computations

Many astronomy laboratories provide small telescopes for use by the students during observing labs. Obtain the objective diameter and focal length of each telescope model from your instructor or from labels attached to the telescopes. Some common eyepiece focal lengths are 5, 10, 15, 20, 25, and 40 mm.

1. Spend some time examining the telescopes in the laboratory. Learn the locations and functions of the controls so that you may operate the telescopes in dark conditions.

2. Make a table containing the eyepiece focal lengths available in the laboratory.

3. Append columns to the above table containing the magnifications produced by each eyepiece with each telescope.

4. Determine the maximum effective magnification of each telescope using the 20X per cm rule. Place these values in the table under the appropriate telescope columns. Do any eyepieces used in the laboratory exceed the maximum effective magnification of any telescopes? If so, indicate which eyepieces and which telescopes.

5. Compute the f/ratio of each telescope.

Telescopes II

Optical telescopes have allowed us to see details and objects never before imagined. But the views presented by telescopes vary according to how the light is collected and focused. Even telescope mounts vary widely. This lab explores various types of telescope optics, mounts, and accessories. More information on telescope design and use can be found in Exercise 8, Telescopes I.

Light can be collected and focused by refraction or reflection. **Refractors** are telescopes that use a converging lens to collect and focus light. The astronomical refractor and the Galilean telescope are examples of refractors. **Reflectors** are telescopes that collect and focus light using a concave, curved mirror. The Newtonian telescope is a reflector design that is still popular today. **Catadioptric** telescopes use both mirrors (catoptric) and lenses (dioptric) to collect and focus light. The Schmidt-Cassegrain telescope (SCT) is a popular catadioptric. Whatever design is used, the primary focusing element (lens or mirror or both) is made large so as to collect as much light as possible. The image produced by the primary focusing element is viewed with an eyepiece, which produces a magnified image. See Exercise 8, Telescopes I, for more information on telescope optics.

The optical geometry of the telescope may cause the image to be inverted or reversed. See Figure 9-1 for examples of these view orientations. The astronomical refractor, for example, produces an inverted image. It might be argued that this is a disadvantage of the telescope. But, Earth-biased ideas of up and down are often irrelevant in astronomy. An astronomical telescope can be converted to terrestrial use with an erector lens or a system of prisms, as is done in most binoculars.

Telescopes may be mounted in a variety of ways. Two common mounting schemes are **altitude-azimuth** and **equatorial**. An **alt-az** mount allows the telescope to rotate parallel to the horizon (in azimuth) and perpendicular to the horizon (in altitude). The **traditional alt-az mount** consists of a U-shaped yoke atop a tripod, see Figure 9-2. The yoke may rotate freely about a vertical axis. Bolts on the tongs of the yoke grip the telescope in its midsection, allowing the telescope to rotate about a horizontal axis. Traditional alt-az mounts are inexpensive, require no special alignment before use, and are easy to aim at objects on Earth. For these reasons, they are commonly used with cameras and small terrestrial telescopes. One problem with a telescope using a traditional alt-az mount is that it can not be aimed at the zenith, as the yoke gets in the way.

The **Dobsonian mount** was designed by John Dobson as an inexpensive, alt-az mount for large reflectors. It is known for its intuitive operation, ease of aiming, and smooth motion. Figure 9-3 shows a Newtonian telescope in a Dobsonian mount. The Dobsonian alt-az mount consists of two major parts, a platform and a box. The box is generally rectangular with the smaller sides forming the top and bottom, and the top and one of the sides is open. Also, the side opposite the open one is less than half the height of the box.

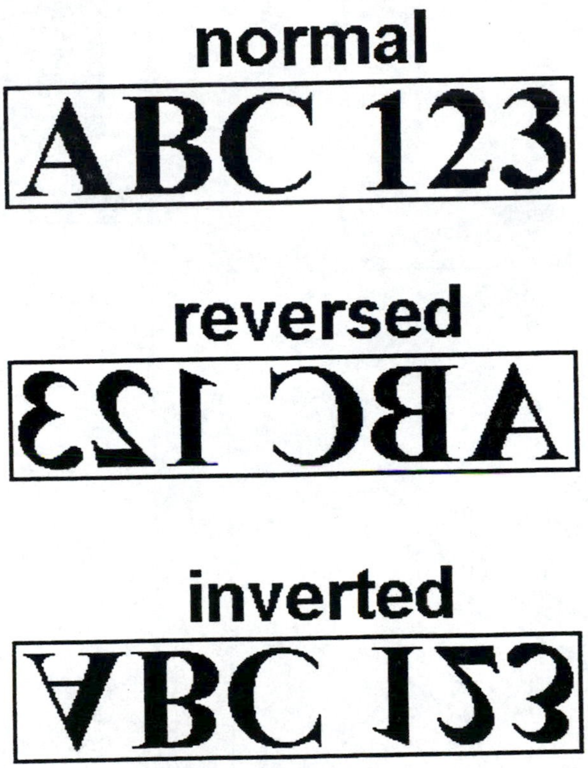

normal

ABC 123

reversed

ABC 123

inverted

ABC 123

Figure 9-1: View orientations.

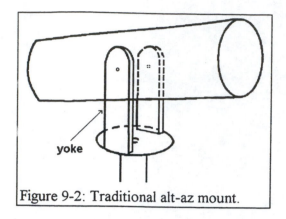

Figure 9-2: Traditional alt-az mount.

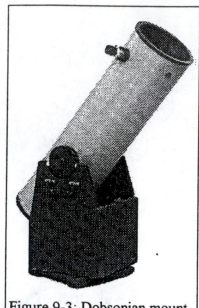

Figure 9-3: Dobsonian mount.

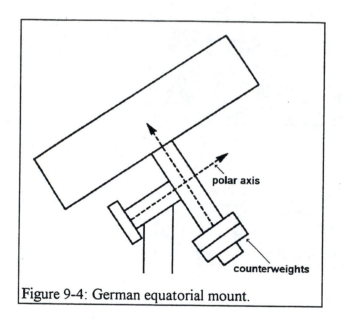

Figure 9-4: German equatorial mount.

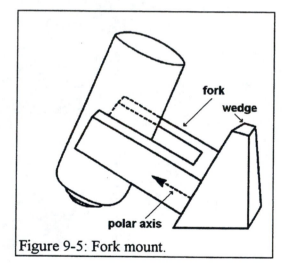

Figure 9-5: Fork mount.

This allows the telescope to tilt in that direction. The bottom of the box is attached to the platform by a bolt through its center. This allows the box to rotate 360° in azimuth. The two other sides have semi-circular cutouts at their tops. Circular bearings are attached to the telescope just farther from the primary mirror than the telescope's center of gravity. The telescope is placed into the box with its bearings resting in the semi-circular cutouts. The box must be tall enough to accommodate the length of the telescope from the primary mirror to the bearings. The telescope is rotated in altitude on these bearings. This variation of the alt-az scheme allows the telescope to view objects at the zenith.

There are several varieties of equatorial mounts. The **traditional equatorial mount** (also called the **German equatorial mount**) has a yoke like the traditional alt-az mount, but the yoke is tilted so that its tongs point toward the celestial equator. See Figure 9-4. The angle between the celestial equator and zenith is equivalent to the latitude where the telescope is used. Thus, the tilt of the yoke must be adjusted for different latitudes. Since the yoke is tilted out of the way, this type of mount allows observers to view objects at the zenith. Rotating the yoke swings the telescope parallel to the celestial equator (in right ascension). Rotating the telescope up and down in the yoke moves the telescope in declination. The traditional equatorial mount requires counterweights to keep the telescope balanced.

The **fork mount** is another equatorial mount design. The fork (or yoke) is mounted on a wedge as shown in Figure 9-5. The slope of the wedge holds the tongs of the fork pointing toward the north celestial pole (or toward the south celestial pole in the southern hemisphere). Rotating the fork on the surface of the wedge rotates the telescope in right ascension. Declination is accomplished by rotating the telescope in the fork. This design is most often used with Schmidt-Cassegrain telescopes.

Telescopes often are used in astrophotography, where it may be necessary to track the objects as they are being photographed. As Earth rotates, a star will rise in the east and travel parallel to the celestial equator to set in the west. A star that does not pass through the zenith, however, changes both its altitude and its azimuth at different rates as it crosses the sky. An alt-az mount thus requires two variable speed motors to follow a star. The traditional alt-az mount has the additional difficulty of not being able to point to objects near the zenith. The Dobsonian mount is able to aim at the zenith, but also requires two motors. Equatorial mounts overcome both of these difficulties.

Equatorially mounted telescopes can track stars with a single motor that drives its right ascension movement at the rate of Earth's rotation. However, equatorial mounts must be aligned with the rotational axis of Earth if they are to track stars correctly. This is called **polar alignment** and may require some effort, but need only be done once for permanently mounted telescopes.

Eyepieces usually consist of multiple glass lenses (or **elements**) and, when combined with the primary optics of a telescope, produce magnified images of the objects being viewed. An eyepiece is usually specified by its focal length, but it is also known by other characteristics, such as eye relief, field of view, the number and types of elements, and casing size. **Eye relief** (or **exit pupil**) refers to the distance from the top element of the eyepiece to where the relaxed eye should be placed to see a clear image. A short eye relief tires the eye quickly, because the lens of the eye must work to produce a clear image. If the eye relief is too long, it becomes difficult to find a comfortable position for the eye.

An eyepiece also is rated for the field of view (measured in degrees) that it provides. A wide field of view shows more of the sky than a narrow field of view. The eyepiece casing comes in three standard sizes 2.45 cm (0.965"), 3.18 cm (1.25"), and 5.1 cm (2"). Most amateur telescopes use 3.18 cm diameter eyepieces. Typically, the cost of an eyepiece increases with the field width, the casing size, and the number of elements.

There are several popular types of eyepieces, which vary in the type of glass used and the number of elements. The **Huygens** and **Ramsden** eyepieces are relatively inexpensive, 2-element eyepieces. Huygens eyepieces have short eye reliefs (typically, around 3 mm) and are uncomfortable for long observing sessions. The Ramsdens have longer eye reliefs (around 10 mm) and are less tiring. These eyepieces most commonly appear on small, department store telescopes. Huygens and Ramsdens have small fields of view (around 30°).

The **Kellner** is a 3-element eyepiece and provides sharp, bright images for low to medium magnifications. The eye relief of the Kellners is between that of the Huygens and Ramsdens, and fields of view are around 40°. Kellners provide good views of the Moon and planets with large apparent sizes. The 4-element **Orthoscopic** eyepieces have long eye reliefs (about 20 mm) and provide excellent sharpness and moderately wide fields of view (around 45°). "Orthos" are less expensive eyepieces for general observing, including lunar and planetary studies.

Four-element **Plössl** eyepieces have fields of view of about 50° and provide sharp images even near the edges of the field. Plössl eye reliefs are long, like "Orthos," so they are comfortable to use for long periods of time. Plössls are excellent for planetary, lunar, and dark sky observing. **Erfles** are 6-element eyepieces optimized for wide fields of view (around 60°), and work best at low to medium magnifications. This makes Erfles excellent for viewing dark sky objects (like star clusters, nebulae, and galaxies), which are usually large in the sky but very dim. Erfles have long eye reliefs like "Orthos." There are many other eyepiece designs. For example, the **Ultra-wide types** use six to eight elements, and provide fields of view so large that the observer must move the eye around to see everything.

A common accessory lens is the **Barlow lens**. One end of the Barlow lens is placed where an ordinary eyepiece is inserted into the telescope. The eyepiece then is placed in the other end of the Barlow. This has the effect of increasing the magnification provided by the eyepiece by a factor of 1.5 to 3 times. However, light now must pass through additional layers of glass, which dims the image.

Procedures

Apparatus
various types of telescopes, mounts, eyepieces, and accessories,
lamp dimmer with clear-glass bulb.

A. Examining the Telescopes
1. Examine the telescopes available in the laboratory. Place their brand and model names in a column of a table.

2. For each telescope model in your table, record whether the telescope is a refractor, reflector, or catadioptric.

3. Note the type of mount each telescope has and place this information in your table.

4. In your table, record the objective focal length and objective aperture diameter of each telescope.

B. Examining the Eyepieces
1. List the types of eyepieces available in the laboratory.

2. List the focal lengths of the available eyepieces.

3. List any specialty lenses with their specifications (e.g., 2X Barlow lens).

C. Viewing through the Telescopes
Set up a telescope of each available model. Use the lowest magnification available (i.e., the eyepiece with the longest available focal length). Aim and focus the telescope on a brightly illuminated object, such as a sign or poster. Answer the following questions for each model of telescope. Record your results in a table.

1. Is the view upright or inverted? Is the view reversed? Use Figure 1 to help you decide.

2. While looking through the telescope, move the objective end of the telescope to the left. Which way does the image appear to move?

3. Which way does the image appear to move when the objective end of the telescope is lowered?

D. Computing Magnification

1. For each telescope model, compute the maximum effective magnification. Arrange these values in a table. The maximum effective magnification of a telescope is given by the 20X per centimeter rule,

$$M_{max} = \frac{20X}{cm} \cdot D,$$

(1)

where D is the objective diameter of the telescope in centimeters.

2. Choose three eyepieces from those listed in part B. One should have the shortest focal length, one should have the median focal length, and one should have the longest focal length.

3. For each available telescope model, compute the magnification produced by each of the three eyepieces chosen above. Magnification is given by

$$M = \frac{f_o}{f_e},$$

(2)

where f_o is the objective focal length of the telescope and f_e is the eyepiece focal length.

4. Do any of these eyepiece-telescope combinations produce magnifications greater than the maximum effective magnification of the telescope? If so, which telescopes and which eyepieces?

5. Try each of the three eyepieces with a Barlow lens. Calculate the new magnifications produced by the telescope system. Compare the quality of the images with and without the Barlow lens.

E. Measuring Objective Focal Lengths

1. Place a screen (or white piece of cardboard or paper) next to an illuminated lamp with a clear-glass bulb.

2. Aim the objective end of a reflector telescope at the screen and move the telescope to focus the image of the bulb's filament on the screen.

3. The distance from the lamp filament to the primary mirror is the object distance (o) and the distance from the screen to the primary mirror is the image distance (i). Use the lens formula,

$$\frac{1}{f} = \frac{1}{o} + \frac{1}{i},$$

(3)

to find the focal length of the primary mirror.

Colors and Spectra

It is widely appreciated that color provides useful information about many visual objects. Scientists have developed many devices for analyzing the spectral characteristics of light, which are referred to collectively as spectrometers. This exercise explores the wave nature of light. Since human eyes will be used as detectors, additional information about the performance of eyes will be acquired. Spectral analysis is so important in astronomy that it occupies the majority of large telescope observing time.

Diffraction gratings are used in these exercises to separate light into its component colors. A diffraction grating consists of many parallel slits, a distance d apart, arranged in a plane. These gratings will be mounted so incident light is normal to the plane of the grating. The Dutch scientist Christiaan Huygens (1629-1695) noted that when a wave is incident on a barrier with perforations, each perforation produces its own wavelet with the same wavelength and speed as the parent wave. As each wavelet spreads out from the barrier, it runs into wavelets from other perforations. The interference between these wavelets produces the diffraction patterns described below.

Consider Figure 10-1. When monochromatic (one wavelength) light passes straight through the grating (where the angle θ is zero) and arrives at the screen (or at the viewer's retina) in phase. The light will be bright there (i.e., light can be seen). To either side of this undiffracted beam of light, the waves from different slits must travel different distances to reach the same point on the screen (or retina). The difference in light path length from adjacent slits, δ, is

$$\delta = d \cdot \sin\theta.$$ (1)

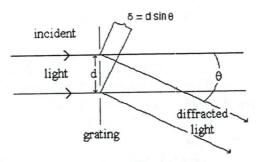

Figure 10-1: Light passing through two adjacent slits in a diffraction grating.

When δ is an integral multiple of the light's wavelength, then the wavelets augment each other. This is **constructive interference**, which is described mathematically by the bright-line formula:

$$n\lambda = d \cdot \sin\theta, \tag{2}$$

where n is the order of diffraction (a positive integer) and λ is the wavelength of the light. First order (n = 1) beams form at the same angle on both sides of the undiffracted beam. Light is also reflected from the diffraction grating, producing first order beams back toward the light source. Thus, there are four first order beams. Other beams are formed for larger values of n, as shown in Figure 10-2.

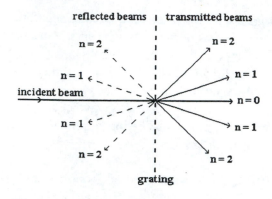

Figure 10-2: The diffracted beams of monochromatic
light incident on a diffraction grating.

Destructive interference occurs when interfering waves cancel each other. With a diffraction grating of many slits, this occurs for all angles other than those for which constructive interference occurs.

Example 1: If a laser beam of wavelength 640. nm passes through a diffraction grating with an interslit distance of 1700. nm, it produces several beams at angles where the bright-line formula, Equation 2, is satisfied. Table 1 below lists the values of n and θ for which beams may be seen using this setup. The brightest diffracted beams occur for n equals one, with progressively dimmer beams for higher orders. Notice that when nλ is greater than d, the sine function is undefined. This means that no higher orders of beams are produced.

Table 1: Angles of diffracted beams
when λ = 650. nm and d = 1700. nm.

n	θ
0	0.0°
1	22.5°
2	49.9°
3	------

When polychromatic (many wavelengths) light is used, each wavelength is diffracted at a different angle, giving a spectrum of colors for each value of n.

Example 2: Table 2 below lists the angles at which blue (450. nm), green (550. nm), and red (650. nm) light are diffracted for the first three orders when the interslit distance is 1700. nm. For this interslit distance, there are no red beams of the third or higher order. Similarly, there are no green or blue beams of the fourth or higher order.

Table 2: Angles of diffracted beams for different wavelengths.
d = 1700. nm

λ (nm)	θ (n = 1)	θ (n = 2)	θ (n = 3)
450	15.3°	32.0°	52.6°
550	18.9°	40.3°	76.1°
650	22.5°	49.9°	------

Wien's law relates the color of an incandescent object to its temperature. The law states that the wavelength of the brightest color, λ_{max}, is inversely proportional to the temperature T, or

$$\lambda_{max} \propto \frac{1}{T}. \tag{3}$$

λ_{max} is at the peak of the **radiation curve**, a plot of light intensity versus wavelength, for an object at temperature T. An idealized radiation curve for a given temperature is called a **blackbody curve**. Figure 10-4 shows some blackbody curves for different temperatures. The radiation curves of stars closely resemble blackbody curves. See Exercise 34, Blackbody Radiation, for more information.

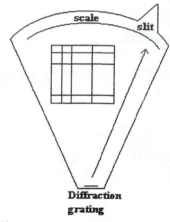

Figure 10-3: A hand-held spectrometer.

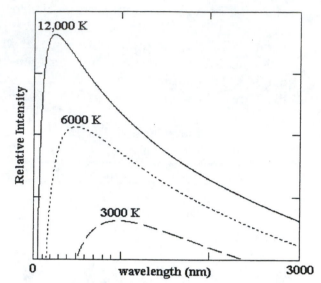

Figure 10-4: Blackbody curves for some different temperatures.

Example 4: One star has a temperature of 6000 K (T_1) and produces its brightest color at a wavelength of 483 nm (λ_{max1}). What is the brightest wavelength (λ_{max2}) produced by a star at a temperature of 3000 K (T_2)?

Using a ratio,

$$\frac{\lambda_{max2}}{\lambda_{max1}} = \frac{1/T_2}{1/T_1} = \frac{T_1}{T_2} = \frac{6000 \cdot K}{3000 \cdot K} = 2.$$

Thus, the peak wavelength is doubled when the temperature is halved. The 3000 K star emits its brightest light in the infrared portion of the spectrum ($\lambda_{max2} = 2 * \lambda_{max1} = 966$ nm). The radiation curve of this star peaks in the infrared, but it also produces light of both longer and shorter wavelengths at lower intensity. Most of the visible light from the 3000 K star is in the red portion of the visible spectrum, so this star appears red to the human eye. Similarly, the radiation from the 6000 K star peaks in the blue, but it also produces light throughout the visible portion of the spectrum with high intensity, so it appears yellow-white to the human eye.

The **Stefan-Boltzmann law** states that the power, P, emitted per unit of area of an object is proportional to the fourth power of its temperature, T. Mathematically, this is

$$P \propto T^4 . \tag{4}$$

Luminosity is the total power emitted by an object. The luminosity of an object increases with the fourth power of the temperature, according to the Stefan-Boltzmann law, and also with the object's surface area. Luminosity is not a property of the object's distance, and to emphasize this, it is often called **absolute luminosity**. The **apparent brightness** of an object depends on its absolute luminosity, the portion of the spectrum detected by the observer, and its distance from the observer. As seen with Wien's law, a star at a temperature of 30,000 K would emit a large amount of its light as ultraviolet. Such a star would appear deceptively dim to the human eye for which ultraviolet light is invisible.

Example 3: Consider two stars of the same size, but one at a temperature of 6000 K and the other at a temperature of 3000 K. How many times more luminous is the hotter star?

Since the stars are the same size, the luminosity ratio is the same as the power per unit area ratio. Thus,

$$\frac{P_1}{P_2} = \left(\frac{T_1}{T_2}\right)^4 = \left(\frac{6000 \cdot K}{3000 \cdot K}\right)^4 = (2)^4 = 16 \approx 20 .$$

Apparent brightness, B, decreases with the square of the distance, D, of the object from the observer. Mathematically,

$$B \propto \frac{1}{D^2} . \tag{5}$$

Example 4: Consider two stars of the same luminosity, but one 10 times farther away from the observer. How many times brighter does the closer star appear than the farther star? Since the apparent brightness decreases with the distance squared,

$$\frac{B_c}{B_f} = \frac{1/D_c^2}{1/D_f^2} = \left(\frac{D_f}{D_c}\right)^2 = (10)^2 = 100$$

The closer star would appear 100 times brighter.

Procedures

Apparatus

hand-held spectrometer, color filter set (at least red, green, and blue), transmission graphs for each color filter, lamp dimmer with clear incandescent bulb, wooden blocks, and He-Ne laser, diffraction grating (~ 600 grooves per mm).

A. Observing the Spectra from an Incandescent Lamp

The hand-held spectrometer has three principal parts: a collimator slit, a diffraction grating, and a scale. See Figure 10-3 for a rough schematic of a hand-held spectrometer. The collimator slit and scale are at one end of the device. Light falling on this end illuminates both the scale and the collimator slit. The light from the slit passes through the diffraction grating where it spreads into a spectrum. When the user looks into the diffraction grating, the image of the scale is superimposed with the spectrum, and the wavelengths of the spectrum can then be measured.

Arrange the spectrometer on blocks so that light from the middle portion of an incandescent filament passes through the collimator slit of the spectrometer. The spectrometer should not touch the bulb. In a dimly lit room and with the lamp turned on at medium brightness, look through the spectrometer eyepiece.

1. Observe the first order spectrum from an incandescent light source. Create a table giving the wavelengths of the centers of what you judge to be the following colors: violet, blue, green, yellow, orange, and red. Also give the approximate ends of the visible spectrum; these are the ultraviolet edge and the infrared edge. Note that different observers may see colors somewhat differently. Also, due to parallax, small movements of the observer's head may affect the measurements.

2. Tungsten filament lamps typically operate below 3000 K, as the melting point of tungsten is 3680 K. Slowly cool the incandescent lamp, and note that the blue end of the spectrum disappears before the red end. According to Wien's law, a cooler object emits less blue light than a warmer object. The Sun's surface temperature is about 6000 K. How would you expect the spectrum of the Sun to compare to the spectrum of the lamp? Which is hotter, the surface temperature of a blue star of the Pleiades cluster or the surface temperature of the Sun? Betelgeuse (Alpha Orionis) has a surface temperature of approximately 3100 K. What is the most likely color of Betelgeuse?

B. The Transmission Spectra of Optical Color Filters

Often a manufacturer of color filters provides a transmission graph for each filter to indicate which wavelengths are transmitted and which are blocked. Typically, the transmission curves are plotted with the percentage of transmitted light on the vertical axis and the wavelength of the light on the horizontal axis. Examine a set of transmission graphs. Notice that all colors (wavelengths) are not transmitted equally.

1. With a bright incandescent lamp, use a spectrometer to determine the approximate wavelengths of each end of the band of light passed by the red, green, and blue color filters. Record these values in a table.

2. Determine the published values using the transmission curves provided by the manufacturer. Record these values in the above table.

3. Do these numbers agree with the manufacturer's values? If they do not, compute the percentage errors and give some reasons why your observations do not match the manufacturer's data.

C. Diffraction of Light

A diffraction grating is rated by its interslit distance, d. Sometimes, the diffraction grating is described by the number of grooves per some unit of length, such as "6000 grooves per cm." This description is the reciprocal of the interslit distance. Thus,

$$d = \frac{1 \cdot cm}{6000.} = \frac{10^{-2} \cdot m}{6000.} = 1.667 \cdot 10^{-6} \cdot m = 1667 \cdot nm$$

Warning: Do NOT look directly into the laser or allow the laser beam to enter your eye directly.

1. Setup the He-Ne laser at one end of the lab. Place the diffraction grating into the beam of the laser. Diffracted beams will be both transmitted through the grating and also reflected back toward the laser. Angle the diffraction grating as directed by the instructor. Describe what you see. How many orders of diffraction do you see? The peak wavelength of He-Ne lasers is typically 633 nm. Record the interslit distance of the diffraction grating. Explain why the number of orders is limited to the number you saw by using the bright-line formula. Hint: Determine for what values of n the formula is valid in this experiment.

2. If a diffraction grating has 6200 grooves per centimeter, what is its interslit distance?

3. Using the diffraction grating of C.2, determine the angles at which a laser beam of wavelength 550 nm would be diffracted. Arrange all values in tables. Sketch all transmitted beams, showing the angles of diffraction.

4. If a diffraction grating with an interslit distance of 1750 nm were used, determine the angles at which 650 nm, 550 nm, and 450 nm light would be diffracted. What are the colors of light with these wavelengths? In the same sketch, draw all transmitted, first-order beams. Arrange all values in tables.

5. What is the largest order, n, produced by the diffraction grating in C.4 for 550 nm light?

Atomic Spectra

The analysis of atomic and molecular spectra allows astronomers to know the compositions and physical conditions of distant celestial bodies. This lab considers atomic structure and the basic principles of how the compositions of both remote and local objects can be determined through spectral analysis.

A **continuous spectrum** is familiar as the smooth progression of colors in a rainbow. See Exercise 10, Colors and Spectra. The bending of light as it passes from one substance to another is called **refraction**. The angle at which the light bends depends on the wavelength of the light and properties of the material. Light of longer wavelengths, such as red, refracts less than light of shorter wavelengths, such as blue. Thus, when white light enters a drop of water, it spreads out into the spectrum of colors. A prism is a piece of glass that is used to spread out the spectrum of light by refraction. **Diffraction** gratings rely on the interference of light waves to spread light into its constituent colors. However, shorter wavelengths diffract less than longer wavelengths, so the spectrum appears in reverse order from that produced by refraction. See Exercise 14, Diffraction and Interference.

Some sources of light, such as fluorescent and "neon" lamps, produce **discrete spectra**, where all colors are not produced. Discrete spectra come in two general varieties: emission and absorption. An **emission spectrum** is produced when a dilute gas is heated to the plasma state. Under these conditions, only certain characteristic wavelengths of light are emitted. When this light is sent through a prism or diffraction grating, the resulting spectrum contains a few bright lines on a dark background. Since the emitted wavelengths are particular to the atomic elements in the plasma, the composition of the substance can be determined.

If the density of the plasma is increased, the lines smear, forming continuous bands. At high enough densities, the spectrum becomes continuous. A "low-pressure" sodium lamp contains a plasma of sodium at about 10^{-5} atmospheres of pressure and produces only two bright, yellow lines at 589 and 590 nanometers. A "high-pressure" sodium lamp contains a plasma of sodium at almost an atmosphere of pressure and nearly produces a continuous spectrum with only a few missing bands. Some high-pressure lamps operate at 10 atmospheres.

Light that passes through or reflects off a substance can be analyzed by the wavelengths absorbed. White light, which consists of all the colors of the rainbow, can be sent through a gas, where certain wavelengths characteristic to the atoms and molecules of the gas will be absorbed. The spectrum that is seen on the other side of the gas would be continuous except for the missing, absorbed wavelengths. This is called an **absorption spectrum**, and can be used to determine the composition of the gas.

A mixture of elements shows the spectral lines of each element in the mixture. This allows astronomers to determine the compositions of distant celestial bodies by comparing their spectra with the spectra of substances on Earth. The intensity of the lines provides information on the relative abundances of substances in the body, its pressure, and its temperature. Also, the Doppler effect shifts the lines by an amount that depends on the relative motion of the body. Approaching objects have their lines shifted to shorter (bluer) wavelengths, while objects moving away have their lines shifted to longer (redder) wavelengths. Doppler shifting of light from the rings of Saturn, for example, shows that the rings are rotating around the planet in the same direction the planet rotates. The rings on one side of the planet show spectral lines that are shifted to shorter wavelengths (coming toward the observer), while spectral lines of the rings on the other side are shifted to longer wavelengths (moving away from the observer).

Spectroscopy is an extremely important part of astronomy. It is also important in other sciences and technologies, such as chemistry and forensics. Tables are available in handbooks that list spectral lines for many elements and compounds. Lists of the "Persistent Lines of the Elements" are usually arranged by element in one place and by wavelength in another. This allows one to look up the wavelengths of the spectral lines of an unknown substance, such as might be found at a crime scene, and determine its composition. The number and the intensity (but not the wavelength) of the spectral lines vary somewhat with experimental conditions, such as temperature and pressure. Thus, it may not always be possible to see all of the lines for a particular element.

The energy of a quantum of light (a photon) is inversely proportional to its wavelength. Thus, a photon of blue light has higher energy than a photon of red light. In equation form, this relation is given by

$$E = \frac{hc}{\lambda},$$ (1)

where h is Planck's constant ($6.63 \cdot 10^{-34}$ joule·seconds) and c is the speed of light ($3.00 \cdot 10^{8}$ meters/second).

The MKS unit of energy is the **joule (J)**. One joule is the amount of energy required to accelerate a one kilogram mass at a rate of one meter per second per second for a distance of a meter or, equivalently, the energy of lifting a one kilogram mass to a height of about 10.2 centimeters on Earth. Another unit that is used when dealing with small amounts of energy is the **electron-Volt (eV)**. An electron-Volt is the amount of energy associated with an electron crossing between the electrodes of a one-Volt battery. The relation between electron-Volts and joules is

$$1 \text{ eV} = 1.602 \cdot 10^{-19} \text{ J}.$$

Example 1: Determine the energy (in joules and electron-Volts) associated with a photon of green light at a wavelength of 550. nanometers.

$$E = \frac{hc}{\lambda} = \frac{\left(6.63 \cdot 10^{-34} \cdot J \cdot s\right)\left(3.00 \cdot 10^{8} \cdot \frac{m}{s}\right)}{550. \cdot 10^{-9} \cdot m} = 3.62 \cdot 10^{-19} \cdot J$$

$$E = \left(3.62 \cdot 10^{-19} \cdot J\right)\left(\frac{1 \cdot eV}{1.602 \cdot 10^{-19} \cdot J}\right) = 2.26 \cdot eV$$

Thus, a photon of wavelength 550. nm has an energy of $3.62 \cdot 10^{-19}$ J or 2.26 eV.

In 1913, Niels Bohr proposed a model of the hydrogen atom now known as the **Bohr model**. In this model, a negatively charged electron orbits the positive nucleus. The energy of the electron determines its orbit: lower energy results in an orbit closer to the nucleus and higher energy results in an orbit farther from the nucleus. The lowest orbit is called the **ground state**. Unlike planets orbiting a star, electrons may reside in only certain discrete orbits, which are unique to each element. Figure 1 illustrates the simplest atom, hydrogen, with one proton in the nucleus and one electron in the ground state. Some other possible orbits for this electron are shown also. This model has been extended to describe higher elements, which have more orbiting electrons and contain both protons and neutrons (electrically neutral particles) in their nuclei.

When an electron falls from a higher orbit to a lower orbit, it releases a photon of light, which carries away the energy difference between the orbits. Since the energy of a quantum of light is inversely proportional to its wavelength, a larger energy difference between the initial and final orbits produces a bluer photon. Referring to Figure 11-1, the light emitted by an electron falling from orbit 3 to orbit 1 is bluer than the light emitted by an electron falling from orbit 2 to orbit 1. In hydrogen, electrons falling to orbit 2 produce visible light. Three of these are listed in Table 1. Since the orbits of each atom are unique, the wavelengths of the emitted light must also be unique. Thus, each atom has its own distinct, spectral "finger-print."

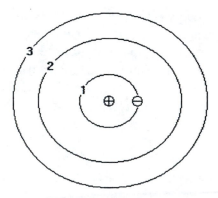

Figure 11-1: The Bohr model of Hydrogen.
An electron is shown orbiting a proton in the
ground state. Two higher orbits are also shown.

Table 1: Wavelengths of prominent spectral lines of some elements in nanometers.
Note: Given colors are approximate.

Helium (nm)	Hydrogen (nm)	Mercury (nm)	Neon (nm)	Sodium (nm)	Approx. Color
		405			violet
447	434	436			violet
469	486		479		blue
502					blue-green
		546	540		green
588		577	585	589	yellow
		579	588	590	yellow
668	656	691	640		red
707			693		red

Consider shining white light on a gas. As the light passes through the gas, some of the photons will "collide" with the electrons of the gas. If a photon has exactly the energy required to boost an electron to one of the valid, higher orbits, then the atom will absorb the photon and the electron will move to the higher orbit. Otherwise, the photon continues on its way. This light produces an absorption spectrum, missing the absorbed wavelengths. The atoms that absorbed photons will emit them in random directions. Thus, light can be seen coming from the side of the gas. This light contains only the emitted wavelengths, and thus, an emission spectrum is seen.

The absorption lines and the emission lines are unique for each element and usually share the same wavelengths. Most molecules are very delicate, and cannot survive being raised to a plasma state. So, a gas of molecules is usually analyzed through absorption. Since the electrons in a molecule are shared between all the atoms making up the molecule, the electrons have many more possible orbits. This adds to the complexity of the energy absorption, and bands of many wavelengths, rather than discrete lines, are absorbed.

Joseph Fraunhofer observed absorption lines in the solar spectrum in 1814, and counted 567 lines of varying intensity, but he was unable to explain their origin. Gustav Kirchoff, in 1859, identified some of these lines as corresponding to the emission lines produced by known elements on Earth. However, by 1868, astronomers realized that some lines did not correspond to any known element. It was thought that there must be elements in the Sun that are not found on Earth. One such element is called helium, after the Greek name for the Sun, *Helios*. Helium was later found on Earth. Sixty-seven distinct elements can be identified from the **Fraunhofer lines** of the solar spectrum. The spectra of the Sun, other stars, nebulae, comets, and other celestial bodies make it clear that the ordinary types of matter found on Earth are widely distributed in the Universe. Spectral studies now provide an important method for classifying stars. See Spectral Classification in the front matter of this manual.

Procedures

Apparatus

hand-held spectrometer, high voltage discharge lamp, gas tubes, incandescent lamp dimmer, incandescent bulb with clear glass and straight filament, low-pressure sodium lamp, sodium absorption tube in oven, spectral chart and lists of the "Persistent Lines of the Elements," and wooden blocks.

A. Calibration of the Spectrometer

1. View the light from the low-pressure sodium lamp through a spectrometer. Identify the approximate wavelengths of the two bright, yellow lines characteristic of a low-density sodium plasma. These lines are very close together and may appear as one line. The two yellow lines were called D by Fraunhofer, and are widely known today as the Sodium D lines.

2. The actual wavelengths of the Sodium D lines are 589. nm and 590. nm. Determine the correction that must be added or subtracted from your readings to determine the true wavelengths of the observed lines. If you saw only one line, you may use 590. nm as your reference.

3. Observe the light from overhead, fluorescent lamps through a spectrometer. These lamps contain mercury. Record the wavelengths, as you read them, of three prominent lines. Find the correct wavelengths by adding or subtracting, as appropriate, the correction for your spectrometer. Compute the percentage error between these corrected wavelengths and the corresponding actual wavelengths listed in Table 1 for mercury. Arrange all your data and results in a table.

B. Emission Spectra

Warning: Typical discharge lamps use about 5,000 volts at about 10 milliamps, which can cause dangerous shocks. Do not handle the lamps when they are turned on. If they need to be moved, turn the power off, move the lamp, and then turn the power back on.

1. Arrange the spectrometers and lamp so that the greatest amount of light from the lamp will pass through the spectrometer's collimator slit.

2. Observe and measure the wavelengths of the three most prominent lines in the spectrum of hydrogen. Record the wavelengths and the approximate colors of these lines in a table.

3. Add or subtract, as appropriate, the correction for your spectrometer to your measurements. Record these corrected wavelengths in another column of your table.

4. List the corresponding wavelengths for the observed lines found in Table 1.

5. Compute the percentage errors between the corrected wavelengths and the published values found in Table 1.

6. Compute the energy (in eV) associated with each photon of the observed lines.

7. Repeat the above activities for helium.

C. Absorption Spectrum

This experiment requires light to be shined through a cloud of sodium gas. However, sodium is solid at room temperature. A tube containing a small amount of solid sodium is placed in an oven where the sodium can be vaporized by raising the temperature to about 200 °C. The oven consists of a box containing heating coils, a temperature control mechanism, and a thermometer.

Light from an incandescent lamp is passed through the tube (and vapor), where light of the wavelengths corresponding to the Sodium D lines is absorbed by the gas. A spectrometer is used to view the light passing through the gas. Two dark lines in the yellow portion of the spectrum can be seen. The dark lines can be "filled in" by scattering light from a low-pressure, sodium lamp off the gas and into the spectrometer.

1. Sketch the apparatus and label its components.

2. Write a brief description and explanation of what you observed. Include a sketch the observed spectrum.

Astrophotography

Photography has contributed extensively to both advancing research in astronomy and bringing discoveries made in astronomy to the attention of the literate world. The gracefully curving arms of giant spiral galaxies, for example, are well known images, yet have been directly observed by very few people. The first known astrophotograph was taken in 1839 when John William Draper captured the Moon on a daguerreotype in a 20-minute exposure. In 1850, William Bond made daguerreotypes of stars. Daguerreotypes were created by one of the first types of photographic processes used, which created images on metal plates. By the beginning of the 20th century, photography was sufficiently advanced that it provided a new method of observing. With improved telescope guidance, astrophotography extended the range of telescopes and also provided accurate images of star fields, which allowed trigonometric parallax to be used to determine the distances to many additional stars.

Astrophotography can increase the range of telescopes because, whereas the eye's sensitivity is not increased by longer observing periods, photographic images build up with increasing exposure times to make dimmer objects visible. Long-period exposures and special, sensitive film extend the range of telescopes by many stellar magnitudes. This allows photographs to be made of dim stars, asteroids, and nebulae that otherwise would not have been seen. Since the mid-1970's electronic devices, such as CCD's, have provided detective performance several times better than photography (see Exercise 13, Electronic Imaging). However, photographic film is still the detector of choice for many applications, because of its low cost, ease of use, and because films are manufactured in much larger formats than electronic detectors. Thus, astrophotography will continue to be an important tool for many years to come.

Astrophotography offers other advantages over direct observing. Special films allow astrophotography to obtain images in the infrared and ultraviolet regions of the spectrum, which cannot be seen directly by the human eye. Astrophotography, as with photography of other subjects, allows the view to be studied again and again, which enables details to be seen that may have been missed in previous viewings or during direct observation. Having a photograph also constitutes a kind of proof of what was observed that can be shared with, and studied by, others. Photographs can also be digitized for storage and analysis on computers.

Disadvantages of astrophotography are that it adds another layer of "technology" between the observer and the observed object, more time and expense are required to see the object, and some films have faults, may be lost or spoiled, and may degrade with time. Also, since photography sums the exposures, variations in star brightness are averaged out and can be lost without leaving a clue of the variations.

In order to understand the advantages and disadvantages of astrophotography it is helpful to understand how photography works, and also how printed black and white pictures are made. A black and white picture is a collection of dots or picture elements, called **pixels**, which collectively cover the picture area. Black areas have black ink on all pixels in them, gray areas have ink on some pixels in them, and white areas have no ink in them so only white paper is exposed. A black and white picture is created by using some method, such as photography, to determine which pixels will be black and which will be white.

Stars observed in the night sky all have the same apparent size, which is determined by the observing instrument used. An unaided human eye can resolve details down to approximately one minute-of-arc. A circular dot 0.1 mm in diameter viewed from 34 cm, a convenient reading distance, subtends one minute-of-arc. Thus, 0.1 mm would be the maximum pixel size for pictures of stars in order to give a true picture of star size, where one pixel with no ink on it would be used to represent each star, and all other areas would be covered with black ink. One problem with such a picture is that all stars pictured would have the same apparent brightness. The only practical way to make some stars brighter than others is to increase the areas representing brighter stars. Photographs conveniently do this, as discussed below, but that makes the pictures inaccurate with regard to a star's apparent size.

Another problem with pictures of stars is that black areas reflect a significant amount of light. The fraction of incident light reflected by an object is called **albedo**. The albedo of paper varies from about 0.7 for white paper to about 0.04 for paper with black ink on it. Thus, if white 0.1 mm pixels are used to represent stars of magnitude 0, then all black pixels of the same size will, by virtue of the light reflected from them, represent stars that are about 17 times dimmer, which would correspond to a star of about magnitude 3. Stars of both 0 and 3 magnitude are readily observable in the sky with the naked eye. Consequently, the light reflected from black areas of a picture will correspond to stars which are dim, but still bright enough to be seen. There is no way in such a picture to distinguish between an area where there is a dim star and an area where there is no star.

Fortunately, human eyes respond to relative differences in brightness and also to edges, so when we see a collection of white pixels in a sea of black, we "see" a field of uniform stars with nothing between. But how can the brighter stars be made to look brighter when all star images in the picture just show parts of the same white sheet of paper? Photography conveniently displays brighter stars more prominently by making them larger. These large star images are technically incorrect, but such pictures are useful. To understand how photography does this, it is necessary to understand some details of the structure of photographic film and how it works.

Photographic film for black and white pictures consists of a clear plastic film that supports a gel layer containing light sensitive materials. Photographic paper has a similar gel layer on white paper. These gel layers contain a large number of tiny crystals of silver halide salts distributed uniformly. The crystals are silver bromide with small amounts of chloride and iodide added. Organic dyes are also included in the gel and adsorb to the crystal surfaces, extending the wavelength range over which the film is sensitive to light. Without dyes, the film would be sensitive only to blue and ultraviolet. Figures 12-1 and 12-2 provide schematic drawings of photographic film.

Making pictures of the stars

Step 1: Expose film

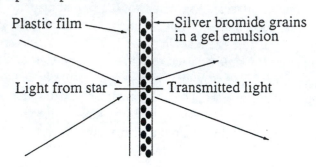

Step 2: Develop film, which makes a negative.

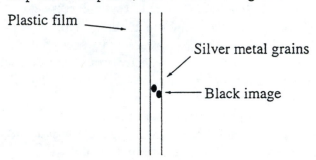

Step 3: Expose and develop a negative of the negative on paper.

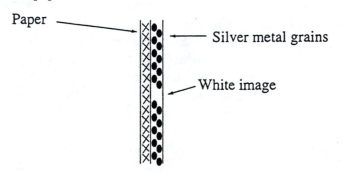

Figure 12-1: The steps in making a picture of a star. In step 1, light from the star is focused on the film, which causes the silver salt grains at that site to develop into silver metal crystals during development, step 2. Development produces a negative, which is transparent except where there are silver metal crystals. In step 3, a positive is made by making a negative of the negative on photographic paper. This makes a positive of the original image.

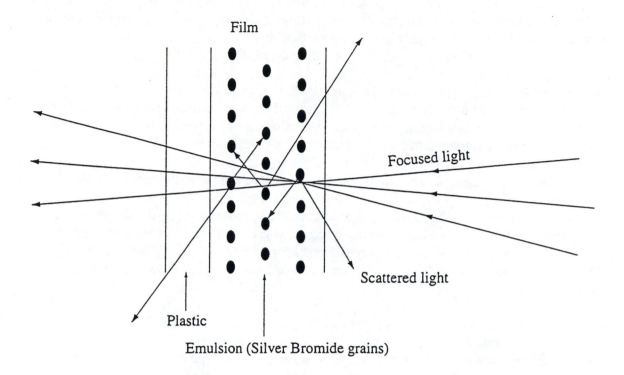

Figure 12-2: A schematic diagram showing photographic film being exposed by star light focused to a point within the gel layer. Some of the photons are absorbed where this light is focused, and on a short exposure, would, after development, produce a small black spot on the negative. As more light is used, with brighter stars and longer exposures, scattered photons will progressively increase the size of the area of the film which will become black during development.

Photographic film must be kept in the dark until it is used. The step in taking a picture when light is allowed to fall on the film is called **exposure**. In most arrangements, exposure uses light from an optical system to create a real image on the film. This light should render the film "developable" so it can be treated in a chemical process, called **development**, that creates a negative image. The film, once developed, is called a **negative**. Sometimes, the negative is used directly for viewing, while other times a second negative is made of the first negative, usually on photographic paper. This second negative is usually called a **positive** (recall that -1 * -1 = +1) or a **print**, and it should look something like the original view.

The part of photographic film that absorbs light to create a picture is the silver salt crystals and adsorbed dye. It is believed that absorbed photons convert silver ions in the silver salt crystal to silver atoms, with a maximum of one atom being converted per absorbed photon. Further, a minimum of two silver atoms must collect together to form a microscopic silver metal grain on the surface of the silver salt crystals to make the crystal "developable." Developable crystals are then converted from silver salt to silver metal during the development process. Thus, two photons must be absorbed by silver salt crystals to constitute exposure. The silver metal crystals that are produced are small, absorb visible light, and therefore appear black. The silver salt crystals not converted to metal in the first part of development are dissolved and rinsed away during a second part of the process. Thus, the film becomes black where light was absorbed during exposure and transparent where it was not. These processes are illustrated in Figure 12-1.

As is illustrated in Figure 12-2, some of the photons incident on the film during exposure are absorbed by the silver salt crystals at that point. But many photons are transmitted and pass through the film, while others are scattered and may be absorbed by nearby crystals. A long exposure would be required to ensure that at least two photons would be absorbed by the silver salt crystals in the image area of a dim star. This long exposure would cause large numbers of photons to arrive at the image points of bright stars in the same field of view. The scattering photons from the bright stars make their images appear larger. Thus, the size of the picture area that develops for stars of different brightnesses is determined by the relative brightness of each star. Hence, the size of a star's image can be measured to determine its relative brightness.

If the telescope moves during exposure, then the astrophotograph will not show dimmer stars, and brighter stars will create lines. The lines created by stars on astrophotographs are called **star trails**. Some astrophotographs are taken specifically to show star trails. Earth's rotation gives stars an apparent motion, which may be captured by time exposures using a camera fixed to Earth. Stars near the celestial equator will move in straight lines while stars nearer to the celestial poles will move in circles centered on one of the poles. For circular star trails, the angular distance the star is observed to move on the photograph can be calculated from Earth's rate of rotation of 15° per hour. Occasionally, a meteor or orbiting satellite may also be photographed with the star trails. Dark sites and dark skies are desired to avoid the film being fogged by stray light.

Astrophotographs that do not show star trails can be obtained by mounting a camera "piggyback" on a telescope that has an equatorial mount and clock drive. Astrophotographs also may be taken by a camera through a telescope. Matching a camera with its lens in place to a telescope with an eyepiece in place is called the **afocal method**. This method provides high magnification, the value of which can be altered by changing the eyepiece, but it is often difficult to accurately align the two devices. The camera body of a single-lens-reflex camera which has a removable lens, may also be mounted on a telescope with a T-ring adapter. In this case the telescope serves as the camera's lens. This places the film at the telescope's prime focus, and so this is the **prime focus** method. The prime focus method provides low magnification and a wide field of view. It works well for photographing the Sun (with an appropriate solar filter), the Moon, and deep sky objects, but the magnification is too low for most planetary photography. Counterweights may be needed on the telescope to avoid overloading the clock drive mechanism, particularly if a heavy camera is used.

Deep sky objects typically require exposure times of 30 minutes or more, during which the telescope must accurately track the target object. Much shorter exposure times are used for photographing the Sun and the Moon, perhaps as short as a millisecond, depending on the telescope aperture, filters, and film speed. Commercial photographic film suitable for astrophotography is available for black-and-white and color negatives, from which prints can be produced, and also color positives, which make slides when mounted.

Procedures

Note: the following procedures include 2 exercises to be performed in a laboratory (A and B) and 4 observing exercises in which astrophotographs are made (C, D, E, and F). Part A explores the images obtained in photographing fiber-star models, and need not be performed before any of the others. Part B familiarizes the students with the equipment to be used in the remaining exercises, and guides the students through planning an observing session, and so should be completed prior to the observing exercises.

A. Photographing Fiber-Star Models
Apparatus

oscilloscope camera, Polaroid film (type 667 or equivalent), fiber-glass star model, and millimeter scale.

1. Load the camera with a film pack, pull the protective black paper from the camera, and turn on the light in the fiber-glass star model. Mount the camera and photograph the star field using a 10 second exposure. Pull the film out according to directions on the camera and allow the picture to develop for 60 seconds. Peal off the negative/developer layer and inspect the print. Label the print so as to identify it as having been taken with a 10 second exposure.

2. Photograph the star model using exposure times that vary by powers of ten, i.e. use 1, 0.1, and 0.01 second exposures, and also 100 and 1000 second exposures. Label each print with the exposure time.

3. Assemble the prints in order of the exposure duration, and describe how the diameter of the fiber-stars' images vary. Measure the diameters of the images from a particular star on each print, and make a table of these measurements along with their exposure times. Plot image diameter versus the logarithm of exposure time (this may be done by graphing the data on semi-log paper, if available.)

4. Comment on the possible usefulness of such a plot for determining the relative brightnesses of a collection of stars that were all photographed in a single exposure.

B. Pre-Astrophotography Session
Apparatus

camera with cable release, film, filters, T-ring adapter, telescope, tripod, and safe-light

1. Examine the camera, telescope and other equipment that will be used. Observe how to operate the telescope's clock drive mechanism. Open the camera body, and observe how the film is to be loaded, advanced, and removed. Attach the camera to the telescope and the cable release to the camera. Adjust the focus and shutter speed. Repeat and discuss these procedures with the instructor and fellow students until they are understood and familiar, as

they will need to be done in semi-darkness when photographing the Moon, planets, stars, and deep sky objects. Familiarize yourself with the operation of the safe light.

2. If solar photography is to be done, inspect and mount the solar filter on the telescope. Note that this filter nearly eliminates the light passing through the telescope in the laboratory. Record the type of light meter that is on the camera, and how it can be read (through the view-finder or on the top of the camera). Plan and discuss the observing exercise with the instructor and class. Record a brief description of your final solar observing plans.

3. If lunar photography is to be done, plan the observing date and time using an almanac or computer planetarium. Determine the rise and set times of the Moon and the location of its terminator on the planned observing day. Discuss the observing plans with the instructor and class, and record a brief description of your final lunar observing plans.

4. If star trail photography is to be done, use a star wheel or computer planetarium to chose two constellations to photograph. A circumpolar constellation and a constellation from the zodiac are recommended. Determine where your target constellations will be in the sky at convenient times as you develop your observing plan. Inspect the camera that will be used, and adjust the f-stop, exposure time (speed), and focus controls through their ranges. Discuss your observing plans with the instructor and class. Record a brief description of your final stellar observing plans.

5. If deep sky objects are to be photographed, determine the target objects and plan the observing exercise. Use a star wheel or computer planetarium to determine where in the sky these objects will be at convenient times, their brightnesses, and rise and set times. Discuss the observing plans with the instructor and class. Record a brief description of your final deep sky observing plans.

6. If planetary photography is to be done, use a star wheel, computer planetarium or almanac to determine the locations in the sky, brightnesses, and angular sizes of the target planets. Also determine their rise and set times. Plan your observing exercise, and discuss your plans with the instructor and class. Record a brief description of your final planetary observing plans.

C. Solar Photography
Apparatus
camera with cable release, film, solar filters, T-ring adapter, telescope, and copies of the film data sheet.

WARNING: do not look through any optical instrument without an approved solar filter in place, or permanent damage to your eyes may result. Also, do not mount a camera on a telescope and aim the telescope at the Sun without an approved solar filter, or the camera may be destroyed.

1. Help set up the equipment. Remove (or block the light path to) the finder scope. Mount the solar filter on the telescope before the telescope is taken into the sunlight. Note in your report, briefly, the activities you performed.

2. Mount the camera on the telescope for prime focus solar astrophotography by taking the lens off the camera and attaching the T-ring adapter. Load film in the camera and attach the camera to the telescope. Attach the cable release and set the shutter speed to the shortest exposure time available. Adjust the telescope for polar alignment, point the telescope at the Sun, and focus the telescope by observing through the camera view finder. Center the image of the Sun in the camera view finder and turn on the clock drive. Record in your report the light meter reading, if it can be read.

3. Record the information requested on the film data sheet. Advance the film and trigger the shutter with the cable release. If this is a new roll of film, repeat this twice more, making three identical exposures and recording that you did this on the film data sheet. This will use the film that may have been exposed while it was being loaded.

4. Adjust the camera to the next longer exposure time (lower speed), advance the film, check the focus, adjusting if necessary, and trigger the shutter again. Record the exposure information in the film data sheet and repeat using longer exposure, until the exposure times are 1/2 second.

5. If afocal astrophotography is to be done, remove the camera from the telescope, replace the eyepiece and camera lens, and mount the camera on the telescope. Center the Sun's image, or images of sunspots if they are in view, in the camera view finder and focus as necessary. Photograph the Sun using the shortest exposure time available on the camera. Record information on this step on the film data sheet. Then check the focus and repeat using longer exposure times. If higher magnification is possible and seems desirable, change the eyepiece, and repeat the appropriate parts of this step.

6. When the solar astrophotography is finished, help put the equipment away. Record, briefly, the activities you performed. Make arrangements with the instructor to have the film developed and prints made.

D. Lunar Photography
Apparatus
camera with cable release, film, filters, T-ring adapter, telescope, tripod, safe-light, and copies of the film data sheet.

1. Help set up the equipment. Note in your report, briefly, the activities you performed.

2. Mount the camera on the telescope for prime focus lunar astrophotography by taking the lens off the camera and attaching the T-ring adapter. Load film in the camera and attach the camera to the telescope. Attach the cable release and set the shutter speed to the shortest exposure time available. Adjust the telescope for polar alignment and point the telescope at

the Moon. Center the Moon in the view finder, turn on the clock drive, and focus on the Moon.

3. Check to see if a light meter reading can be obtained. If it cannot, set the exposure time as short as possible on the camera, advance the film and trigger the shutter with the cable release. Record the information requested on the film data sheet, as well as the light meter reading (if possible). If this is a new roll of film, advance the film and trigger the shutter with the cable release on this same setting twice more, making three identical exposures and recording that you did this on the film data sheet.

4. Adjust the camera to the next longer exposure time (lower speed), advance the film, check the focus, adjusting if necessary, and trigger the shutter again. Record the exposure information on the film data sheet and repeat using longer exposure, until the exposure times are 1 second.

5. If afocal astrophotography is to be done, remove the camera from the telescope, replace the eyepiece, camera lens, and mount the camera on the telescope. Center the Moon's disk or lunar features (such as craters, rilles, mountains, etc.) near the terminator in the camera view finder, and focus as necessary. If light meter readings are not available, photograph these objects using exposure times from 1/250 second to 1/8 second. Record information on each exposure on the film data sheet. Then check the focus and repeat using longer exposure times. If higher magnification is possible and seems desirable, change the eyepiece, and repeat the appropriate parts of this step, using longer exposure times.

6. When the lunar astrophotography is finished, help put the equipment away. Briefly record the activities you performed. Make arrangements with the instructor to have the film developed and prints made.

E. Photographing Star Trails
Apparatus
black paper card, camera capable of timed exposures, cable release, film (400 ASA, for color prints or slides, 24 exposures, recommended), copies of film data sheet, stop watch or watch with second hand, safe light, and tripod.

1. Help set up the equipment. Note in your report, briefly, the activities you performed.

2. Mount the camera to a tripod, set the shutter speed on B ("bulb," for timed exposure), and attach the cable release to the camera. Set the f-stop to its smallest number (so the lens is wide open), and the focus to infinity. Open the camera, load the film, and record the information requested at the top of the film data sheet.

3. Point the camera at the first target constellation. Set the exposure time to 1 minute. Make sure no light source is used close enough to allow stray light to enter the camera. Make a 1 minute exposure. Record information on this exposure on the film data sheet. If this is a new roll of film, advance the film and trigger the shutter with the cable release on

this same setting twice more, making three identical exposures and recording that you did this on the film data sheet.

4. Take several more exposures, varying the exposure lengths as 5 minutes, 10 minutes, 15 minutes, and 20 minutes. Make sure no light source is used closely enough to allow stray light to enter the camera. During the 15 and 20 minute exposures, briefly block the light entering the camera by holding a black card immediately in front of the camera lens for one minute. Start this one minute two minutes after the exposure starts. Blocking the light for this period will create gaps in the star trails which will allow one to determine the direction of star movements in the star trails. Record information on each exposure on the film data sheet. Also, record comments such as "clouds passed by during exposure," "car lights may have shone into camera," and so on.

5. Repeat steps 3 and 4 for the other target constellation.

6. When the star trail photography is finished, help put the equipment away. Record, briefly, the activities you performed. Make arrangements with the instructor to have the film developed and prints made.

F. Planetary Photography
Apparatus

camera with cable release and mounting hardware including a T-ring adapter, film, telescope, safe-light, and copies of the film data sheet.

1. Help set up the equipment. Note in your report, briefly, the activities you performed.

2. Mount the camera on the telescope for afocal astrophotography. Load film in the camera, attach the cable release, and set the shutter speed for 1 second. Adjust the telescope mount for polar alignment, and point the telescope at the target planet. Center the planet in the finder scope and turn on the clock drive. Focus on the target planet through the camera view finder.

3. Make a 1 second exposure. If this is a new roll of film, advance the film and trigger the shutter with the cable release on this same setting twice more, making three identical exposures and recording that you did this on the film data sheet. Check the focus and repeat using longer exposure times. If higher magnification is possible and seems desirable, change the eyepiece and repeat the appropriate parts of this step, using longer exposure times. Record information on each exposure on the film data sheet.

4. Remove the lens of the camera and mount it on the telescope with a T-ring adapter for prime focus astrophotography. Check to see if a light meter reading can be obtained. If it cannot, set the exposure time as short as possible on the camera, advance the film and trigger the shutter with the cable release. Record the information requested on the film data sheet, and the light meter reading (if possible). If this is a new roll of film, advance the film

and trigger the shutter with the cable release on this same setting twice more, making three identical exposures and recording that you did this on the film data sheet.

5. Adjust the camera to the next longer exposure time (lower speed), advance the film, check the focus, adjusting if necessary, and trigger the shutter again. Record the exposure information in the film data sheet. Continue to make longer exposures to bracket the predicted best exposure times for the planet being observed. For Venus and Jupiter exposure times ranging from 1/2 second to several seconds may be ideal. Mars and Saturn will probably require exposures two to four times longer than those required for Venus and Jupiter.

6. When the planetary astrophotography is finished, help put the equipment away. Briefly record the activities you performed. Make arrangements with the instructor to have the film developed and prints made.

G. Report (after film has been developed)

1. List the objects photographed, the camera, the telescope, the methods used, the film type, and the range of exposures used on each target object. Also, record any other relevant information, such as the magnifications and f-ratio of the telescope.

2. Describe your photographic results and describe any difficulties you experienced. For example, was light pollution or telescope vibration a problem?

3. Identify which exposure lengths were best for each subject.

4. If solar photography was done, describe any solar features visible in your photographs.

5. If lunar photography was done, describe any lunar features visible in your photographs.

6. If you made star trails, comment on the star colors. If you made photographs of different parts of the sky, describe why the star trails from different parts of the sky appear differently. Is it possible to determine the direction of the star movement in any of your photographs?

7. If planetary photography was done, describe any planetary features visible in your photographs.

Film Data Sheet for role No. ____   Telescope and camera used: _____

Date: _____   Film type: _____   Observing Conditions: _____

Frame	Exposure	Subject	Notes and Comments
1			
2			
3			
4			
5			
6			
7			
8			
9			
10			
11			
12			
13			
14			
15			
16			
17			
18			
19			
20			
21			
22			
23			
24			
25			
26			
27			
28			
29			
30			
31			
32			
33			
34			
35			

Electronic Imaging

Charge-coupled device (CCD) imaging has begun to make major inroads on traditional, film-based astrophotography. Electronic imaging offers immediate and continuous inspection of observations, as well as a continuous recording of an event or observation. In addition, observers making video recordings can record an audio track during the observation, which may include timing signals from shortwave radio stations or other accurate clocks to accurately time events. This lab explores some of the concerns of electronic imaging of astronomical objects.

Since the 1980's, video cameras and camcorders (video cameras with recorders) have been readily available in a variety of tape formats, from professional to home video. These cameras use a type of light-sensitive electronic chip, which captures images from light that is focused on it. While home video cameras are not ideal for recording low light level events and objects such as deep sky objects or planets, they are excellent for lunar observations, low magnification planetary observations, and lunar and solar eclipses. These home units come with optical systems and need to be adapted for use with telescopes. Professional systems typically put the light-sensitive chip at the prime focus of whatever optical system is used.

Most home camcorders use standard VHS, VHS–C (compact), or 8-mm format. Of the three formats, only the 8-mm format uses metallic tape, which produces a better quality picture. 8-mm metallic tape also suffers less picture degradation over time than the other two formats. However, 8-mm tape can not be played in most VCR's, so it is not as convenient as VHS and VHS-C formats.

Many camcorders include the option of automatic or manual exposure control. Some operate by varying a lens diaphragm (or iris) to control the amount of light falling on the chip, whereas others vary the exposure time. A camcorder's lux rating refers to its light sensitivity. Lux is a unit of illumination, so a lower lux rating means the camera can record images of dimmer objects. Many video cameras advertise lux ratings as low as 1 lux, which is sensitive enough to record many faint details. It is often best to set the exposure control manually. The automatic exposure circuits often are fooled by objects of greatly varying brightness in the field of view, as are usually encountered in astronomy. This tends to "wash out" some of the brighter features or not even show some of the dimmer ones. Likewise, automatic focusing can cause problems too. Since celestial objects are far away, the camcorder should always be manually focused at infinity.

Most camcorders offer a zoom feature. They range from wide angle to telephoto, usually magnifications of 8X, 10X, or 12X. Some models offer a digital zoom feature of up to 24X, but the results are often grainy. Camcorder lens systems usually are not sufficient to record a lunar or solar image of significant size to show detailed structures. Flexibility also is usually restricted by the fact that home camcorder lenses are not easily removed. Often, this is due to the electronic circuits built into the video camera's lens system to control the auto-focus and zoom features.

A telextender is an accessory that can help eliminate image size problems by magnifying the image before it reaches the video chip. Telextenders connect to the front of the video camera's normal lens. They are available in different magnifications, from about 1.5X up to about 5X. Unfortunately, telextenders can degrade the quality of the image. Vignetting, where the image shows curious shading effects, can also result from using a telextender if the camcorder is set to a focal length that is too short.

An alternative to the telextender employs an afocal system. Afocal systems match the camera's lens to the complete optical system (objective and eyepiece) of a telescope. Low power (less than 50X) is recommended; too high a magnification produces a poor image (a washed-out image with vibrations greatly exaggerated, and an overall poor quality). The video camera–telescope system needs to be mounted to some sort of base plate to assure alignment between the components and eliminate any movement of the system. A light block may be needed to eliminate extraneous light that can enter between the components of the system.

Some other features included on some camcorders are useful in astro-imaging. These include time-lapse exposure settings (e.g., one exposure every 30 seconds), date and time labeling, fade-in and fade-out, and subtitling.

There are several options for mounting the camera. These include attaching the video camera to a tripod, piggy-backing it on a telescope, and mounting it directly on an equatorial mount. A standard tripod is useful when the object does not move during the exposure. For long exposures, a moving object may appear smeared across the image if tracking is not provided. A camcorder that is mounted on the tube of a telescope, like a finderscope, is said to be piggy-backed. If the telescope is mounted equatorially and contains drive motors for tracking sky objects, then the piggy-backed camcorder can follow sky objects as Earth rotates. This is especially useful for taking long exposure images. An alternative to this is to place the camcorder on its own equatorial mount with tracking motors.

A CCD is a solid-state light detector. It consists of a silicon chip, which detects incoming photons of light at thousands of individual picture elements (pixels) arrayed in a rectangular grid on a semi-conductor chip surface. The pixels build up an increment of electrical charge each time they are struck by a photon. By measuring the built-up charge, the brightness of the image at each pixel can be determined. These electrical charges are sent to an amplifier and then recorded digitally in a computer or in a magnetic storage medium for later analysis.

One of the many advantages of CCD's over traditional film is sensitivity. The CCD is typically able to record 75 percent of the incoming photons. Traditional photographic methods typically record less than 3 percent of the incoming photons. This fact allows CCD's to detect objects more than 25 times dimmer than photographic film with the same exposure time. In other words, a CCD can show the same level of detail as photographic film in less than 1/25 of the time. Since telescope time is valuable, CCD's allow astronomers to image more objects in less time and to record images too dim to record on film.

Since the images recorded with a CCD are stored digitally, they are in a format that may be easily processed later by a computer. "Noise" in the raw images can be conveniently filtered out at that time. "Noise" is anything that degrades the quality of the image, like backscattered light from the atmosphere. Even optical defects in the telescope and bad "seeing" conditions can be compensated for, at least in part. One of the difficulties with CCD's is the necessarily small field of view. Wide-angle fields are currently beyond the capability of CCD's, but for imaging the shadow of a Galilean satellite on Jupiter's cloud tops or for imaging the Whirlpool Galaxy, the CCD is unparalleled.

Procedures

Apparatus

telescope, video system (this can be as simple as a camcorder and tripod to a telescope-mounted camcorder), several blank video tapes, telextender, monitor, and appropriate interconnecting cables and power supplies.

A. Pre-Observing

1. Examine the video camera in the lab. Note the location of the record button, auto/manual focusing, exposure setting (iris, exposure, etc.), zoom feature.

2. What type of tape does the camcorder require? How long does the battery need to charge? What is the zoom ratio of the camera?

3. Try taking some indoor videos first, familiarizing yourself with the features of the camera. Reduce the lighting in the room and again take some video. How sensitive to the light level is the camera? Practice moving the camcorder around on the tripod.

4. Try the telextender, if available, on the camcorder. What is the magnification of the telextender? Note differences in focusing, image brightness, and image clarity when using the telextender and when not using the telextender.

B. Video Observations

1. The Moon
The Moon makes a great first object for video imaging, as it is bright and easy to find.

    a. Set the camera, on wide-field rather than the zoom part of the telephoto, on its tripod and aim it at the Moon and begin recording. Use the zoom control to increase and decrease the magnification of the system over the full range allowed.

    b. Place the telextender onto the video system, and again record the Moon. Increase the magnification over the full range while imaging the Moon. Do you see any features that you could not see before? How does the light intensity change as the magnification is increased? How does the resolution (clarity) change with increased magnification? You might need to make these observations after returning to the lab and viewing the video tape on a television or monitor.

2. The Sun
The video camera, with its telextender, can also be used to record solar images through appropriate solar filters. All light entering the telextender and camcorder must first pass through an approved solar filter, else the camcorder's detector chip may be destroyed.

a. Record images of the Sun using the full range of magnification.

b. Describe the locations and sizes of any sunspots that are visible. Are there any faculae, plages, filaments, or prominences visible?

3. Eclipses

Use an approved solar filter while observing an eclipse of the Sun. Remember that the Sun and Moon move across the sky during the eclipse, so use appropriate tracking equipment.

a. Note the type of eclipse and how long the eclipse is visible from your site.

b. Use some system for recording the time with the image. This can be done with a date and time labeling system built into the camcorder or by recording shortwave time signals on the audio track.

c. Aim the video system at the Sun or Moon, and record for a few minutes prior to the eclipse. Check the recording to ensure that the system functions properly.

d. Begin recording a few minutes prior to the eclipse and continue recording for a few minutes after the eclipse.

e. What features on the Moon or Sun could you see before the eclipse began?

d. Describe the resolution (clarity) of the eclipse on the video. For a lunar eclipse, how easy was it to see the shadow as it progressed across the Moon's surface? For a solar eclipse, was the Moon's limb sharp as it moved across the Sun?

4. The Planets

The brighter planets can also be imaged without a telescope or even a telextender.

a. Which planet are you observing? In which constellation does this planet currently reside?

b. Aim the video system at the planet and record for a few minutes.

c. About how much of the field of view is filled by the planet at the highest available magnification?

d. Describe the clarity of the image. Also, note any colors, features, etc. that you can see on the video.

e. Note any stars visible on the video. Identify some of these stars.

Astronomy Laboratory Exercise 14
Diffraction and Interference

The performance of optical instruments, including eyes and telescopes, are limited by the diffraction and interference which result from the wave nature of light. This phenomena will be demonstrated in this exercise. The fact that diffraction and interference occurs verifies that light behaves as waves. The diffraction that results from a single linear slit of width w is illustrated schematically in Figure 14-1, where the geometrical shadow is also shown. The first dark band on each side of the central maximum occurs at the angle λ / w (in radians), where λ is the wavelength of the light used. At this angle, light arriving at the screen from the closer half of the slit is half a wavelength or π radians out of phase with light arriving at the screen from the other half of the slit, and thus interferes destructively with it. If light rays traveled in straight lines, only the geometrical shadow of the slit would result. If you are skeptical about such diffraction patterns, make a narrow slit between two fingers, place that slit one centimeter from an eye and look through the slit at a bright light. The multiple edges, alternating light and dark bands seen when the slit appears almost closed, are caused by diffraction. More information on light can be found in Exercise 10, Colors and Spectra.

Diffraction from round holes produces circular bands, with the first dark band occurring with an angle of $1.22 \cdot \lambda / D$, where D is the hole diameter. This value may also be taken as the angular size of the central bright spot, called the **Airy disk** (after an English astronomer, Sir George Airy, 1801-1892, who first explained this phenomenon). Telescopes suffer from diffraction. If the diffraction patterns are large, the images from adjacent stars may overlap and appear as a single star. The diameter of the Airy disk can be reduced by increasing the telescope's objective aperture, which can be taken as the value of D above. Larger apertures make images brighter for two different reasons: because larger diameter telescopes collect more light and because larger diameter telescopes concentrate the collected light into smaller Airy disks.

Diffraction also occurs from the pupil of our eyes, but normally we are not aware of it as our eyes are arranged to minimize diffraction and we tend to ignore it. One might think we could see greater detail in objects if our rod and cone cells were smaller and packed more closely together in our retinae. That would give more **pixels** (picture elements) in a given area, presumably increasing our visual resolution, but we would only see more diffraction. We would need to have larger pupil diameters to take advantage of any higher retinal cell density.

Since the diffraction of light depends on the wavelength of the light, it can be used to separate light into a spectrum. Convenient devices for doing this are **diffraction gratings**, which are flat pieces of glass or plastic with many parallel slits uniformly spaced a distance d apart. Light hitting each slit scatters in all directions and interferes with light from other slits. The paths from different slits to a screen in a particular direction will all be of different lengths. When a laser beam is directed to a diffraction grating, four beams will form at the angle where the path lengths from adjacent slits differ by exactly one wavelength, as shown

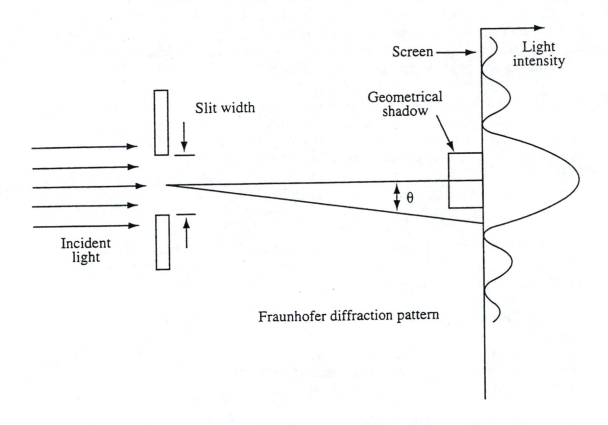

Figure 14-1. A uniform beam of incident light passing through a large slit produces a geometrical shadow. The small amount of diffraction occurring at the edges is small and easy to ignore. So, one might expect the same would result from a narrow slit, but instead a narrow slit will produce a Fraunhofer diffraction pattern, as shown.

in Figure 14-2. That is, $d \cdot \sin\theta = \lambda$. These are identified as first order beams. Higher order beams may also occur where $d \cdot \sin\theta = n\lambda$, where n is the order of diffraction. Destructive interference occurs at other angles.

The term interference is used when light traveling along two paths collide, such as the light from the two surfaces of soap films or from oil films on water. Thin coatings are put on lenses and other optical surfaces to reduce the amount of light reflecting from those surfaces. The amount of light reflecting from a clean air-glass surface is about 4.3%. This means that the fraction of transmitted light will be about 0.957.

Example 1: How much of the incident light would be lost in a device that used two lenses and two prisms, but none of the optical surfaces are coated?
> The fraction of the incident light transmitted through each air-glass surface that is not coated is 0.957. With 8 surfaces, 2 for each lens and prism, the fraction of light passing through all surfaces will be about $(0.957)^8 = 0.704$. Thus, nearly 30% of the incident light would be lost to reflection in this device.

A thin "anti-reflective" coating on glass can cause interference to reduce the amount of reflected light to below 1.3% over the visible spectrum. This increases the amount of light transmitted and produces a brighter image.

Example 2: Find the fraction of light transmitted in example 1 when anti-reflective coatings are applied to the glass surfaces.
> The fraction of light transmitted for each surface increases to 0.987. For 8 surfaces, the fraction of light passing through all surfaces will be increased to $(0.987)^8 = 0.901$.

Typically, such coatings are designed to optimize transmission at a wavelength of 555 nm when it is anticipated that the devices will be used for daylight vision or at 510 nm for night vision. Such coatings do not work as well near the ends of the visible spectrum because of the different wavelengths in those parts of the spectrum, which are red and violet. That is, the coating is not as effective in increasing the transmission of red and violet light. Thus, the relative amount of red and violet light is increased in the reflected light, and this accounts for the purple color seen in reflection from many coated lenses.

Reflections can also create multiple images. This is a particular problem with bright objects such as the Sun. Many photographs or video images taken outside with the camera looking towards the Sun will show multiple images of the Sun. Although coated lenses do not eliminate this problem, they do reduce the brightness of the unwanted images.

A simple and inexpensive coating method is to add one layer of manganese fluoride that is one-fourth of a wavelength thick. This causes the purple color of reflected light, as described above. Multiple coatings can reduce the amount of reflected light even more, and may show other colors in reflection, including the green color seen in popular astronomical

A.

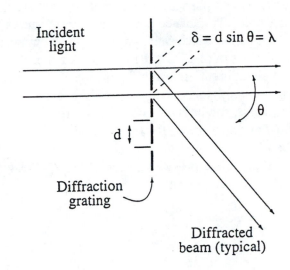

B.

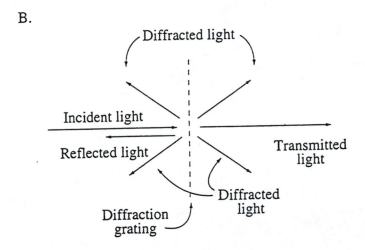

Figure 14-2. A beam of light incident on a diffraction grating will be reflected and transmitted, and will also produce four first order diffracted beams. A. shows how the path length difference from adjacent slits is equal to one wavelength of light. B. shows the four first order beams, two on each side of the grating, which satisfy the relation, $\delta = d \sin\theta = \lambda$.

binoculars. Explanations of how these coatings work requires analysis of the path length differences for reflections from the multiple layers of the coating.

Interference effects in star light also produce twinkling. Bright stars on a clear night often exhibit a color-twinkling phenomenon, sometimes making the image red, sometimes green, and sometimes blue. Turbulence in the atmosphere causes light rays from a star to follow slightly different paths, which causes the rays at one moment to arrive at the eye in phase for one color and out of phase for the others. A moment later, the color of light arriving in phase will be different. In order for the eye to detect this color-twinkling effect, the star must be bright enough to activate cone cells (see Exercise 6, About Your Eyes). When dimmer stars do this, rod cells are activated to produce an on-off twinkling. Planets, typically, do not show either of these effects because their larger angular sizes cause the light rays in the atmosphere to be incoherent, and thus, interference (and twinkling) will not result. The absence of twinkling provides a simple and rapid observational test for distinguishing between stars and planets.

Procedures

Apparatus

aluminum foil, binoculars, bottles (colored glass), calipers, cork, diffraction grating and holder, diffraction slit (0.1 mm or less), interference filters, lamp (tubular with clear glass and straight filament), laser, lenses (coated and un-coated), meter sticks and/or tape measure, microscope slides (sticking together to show interference), paper punch, protractor, spectrometer (hand-held), straight pin, photocopy of Figure 14-3, if possible

WARNING: do not shine laser light into anyone's eyes, or allow laser light to be directed into yours, as direct laser light may damage the retina.

A. Diffraction from Slits

Place a tubular lamp in position where it can be comfortably viewed from several meters distance, with the filament vertical and turned on. Darken the lab. Look at the lamp filament through the diffraction slit provided and note the diffraction pattern. Sketch the diffraction pattern observed, note any colors seen, and show or label the colors in your sketch.

B. Diffraction from Circular Holes

1. Project a laser beam onto a screen across the lab. Measure and record the size of the spot.

2. Then project the laser beam through a pin hole. (A pin hole can be made by placing a piece of aluminum foil against a flat piece of cork and carefully pushing a pin through the aluminum foil and into the cork. The hole diameter can be estimated by measuring the diameter of the pin.) Record the pin hole diameter, d. Measure and record the distance from the hole to the screen, and the diameter of the first dark circular band in the diffraction pattern.

3. Calculate and compare the angular size of the Airy disk to the predicted value of $1.22 \cdot \lambda/d$. λ, for reference, is 633 nm for HeNe lasers. Obtain the published wavelength for your laser from the instructor.

C. Diffraction from Gratings

1. Measure the wavelength of laser light using a grating of known spacing. Fold the paper screens provided as Figure 14-3, punching a hole as shown. Place a diffraction grating vertically in the center of the paper, as shown. Direct a laser beam at the grating through the hole. Observe that at least six beams can be seen, the transmitted and reflected beams, and four first order diffracted beams. Align the grating so the reflected beam is directed back to the laser. Then hold a pencil vertically in one beam so the pencil body interrupts that beam,

and move the pencil along the beam so the pencil point traces the beam path on the paper. Repeat this for the incident, transmitted, and four first order beams. Turn off the laser, then move the grating, and use a straight edge to connect the lines following the beam paths through the position where the grating was. Measure the angles with a protractor. Record these values and find the mean value.

2. Use the formula, $\lambda = d \cdot \sin\theta$, where θ is the average value of the angles, to find the value of λ. Compare your result to the published value for the laser. Alternatively, if the spacing of the grating is not known, but the laser wavelength is, determine the grating spacing.

3. Observe the spectrum of light from the incandescent lamp using the diffraction grating, first by looking directly through the grating at the lamp filament, then looking off to one side, then the other side. You should see the lamp's spectrum on each side. Then, turn so the lamp is behind you and use the grating as a mirror. Rotate the grating mirror to again look off to each side. You should be able to see two reflected spectra, one on each side of the reflected image of the filament. These four spectra correspond to the four first order diffracted beams traced above. Which color is diffracted through the largest angle in each of these beams? Is that expected? Explain why.

4. Use a hand-held spectrometer to look at light from incandescent and overhead fluorescent lamps. The internal scale is useful but may not be very accurate. You should again see the complete spectrum from the incandescent lamp. Sketch the observed spectrum of the overhead lamps. The bright lines result from mercury inside the fluorescent lamps.

D. Interference from Thin Films

1. Microscope slides, when stuck together, typically trap a thin pocket of air between them. The light reflected from the bottom of the top slide will interfere with light reflected from the top of the bottom slide. The air pocket typically varies in thickness as a wedge. This causes the interference pattern to appear as colored stripes. The pattern is easiest to observe in reflected light by holding the slides over a dark object and looking at reflections of overhead lamps. Look for such patterns, sketch them, and note observed colors in your sketch.

2. Interference filters are made using this phenomenon. If interference filers are available, look at the color of the light that is reflected and transmitted by them. Describe your observations.

E. Reflection from Lenses

1. Look at the reflections of the overhead lamps produced by an uncoated converging lens. You should see multiple images. By tilting the lens, you will see the images move. Note that they do not move together. This and other observations should allow you to determine which air-glass surface produces which image. Sketch the images and describe their colors.

2. Repeat the above step using a coated, converging lens (which may be part of binoculars or other optical devices). Compare these results to those with uncoated lenses.

F. Reflections from Colored Glass Bottles

1. Look at the images of overhead or tubular lamps reflecting from the shoulder and neck area of a colored glass bottle. The inside and outside glass surfaces often have different curvatures, which help to separate their respective images. Pick a bottle, sketch and describe the images seen, including their colors and relative brightness. Identify the bottle in your report. Explain how it can be determined which surface is producing the respective images. (Hint: light reflecting from the inside surface must pass through the colored glass twice.)

2. Fill your bottle completely with water, and note any change to the brightness of the images. Does your observation support your explanation given above? Explain how it does, if it does.

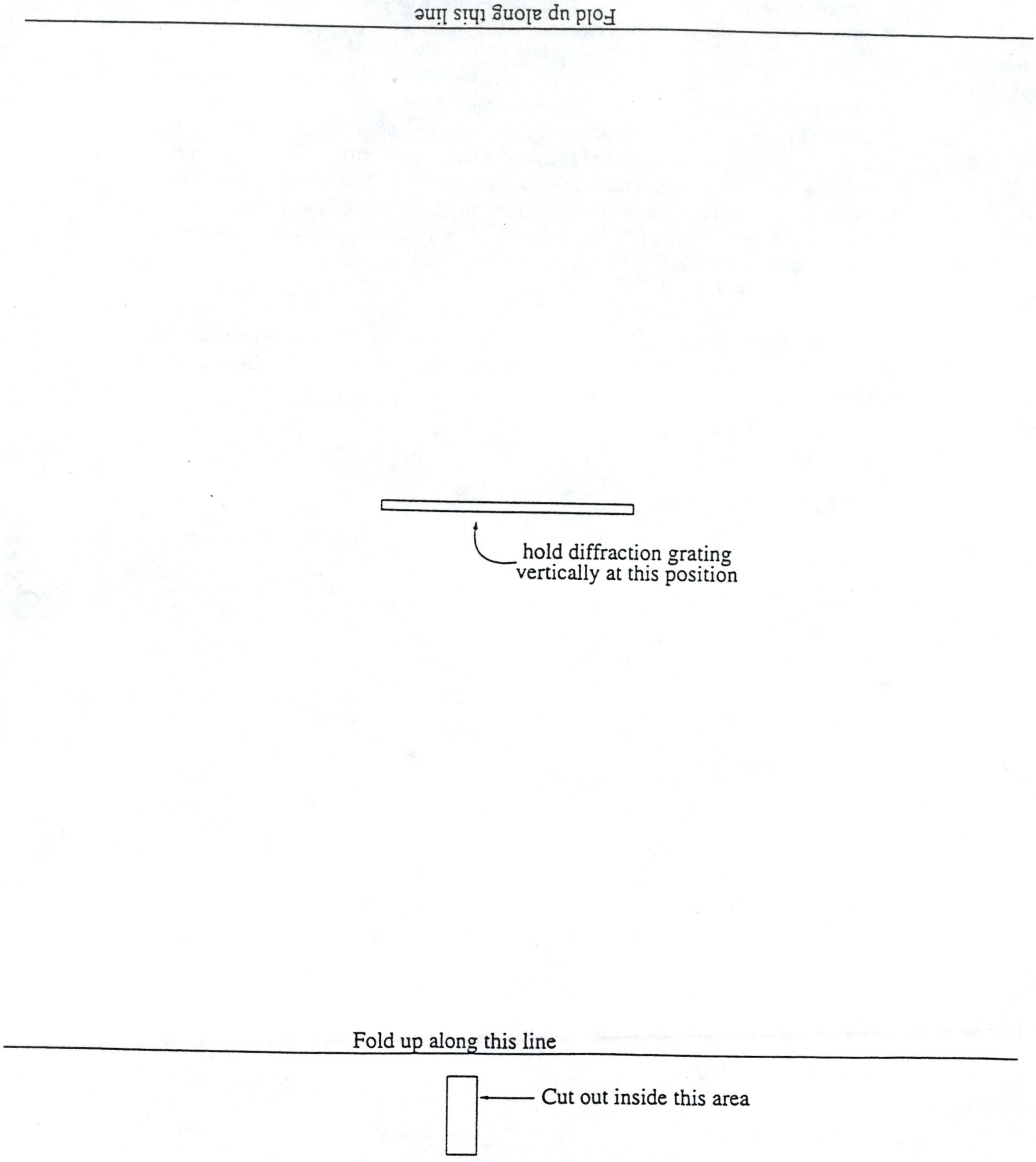

Fold up along this line

hold diffraction grating
vertically at this position

Fold up along this line

—— Cut out inside this area

Figure 14-3. The fold-up screen used to measure of θ, as specified in the procedure.

The Motions of Earth

A significant advantage of the heliocentric model of the Solar System is how it naturally explained the daily and annual motions of the Sun, Moon, and stars as resulting from the rotation and revolution of Earth. But other motions of Earth are required by current models of the Universe, which are less well known. The Earth, for example, moves as a result of the Moon's gravity, as evidenced by tides in the oceans. And Earth moves with the Solar System in an orbit around the disk of the Milky Way galaxy, while the Milky Way orbits in a cluster of galaxies, which are collectively falling toward a much larger cluster of galaxies. These motions of Earth are discussed and compared in this exercise.

Physicists distinguish **velocity** from **speed** by indicating velocity is a **vector**, a quantity having both a magnitude and a direction, while speed is a **scaler**, a quantity having only a magnitude. To illustrate, an automobile's speedometer gives only speed. A driver must know the direction from other information! Speed is measured in units of distance divided by time, such as m/s or km/hr.

Earth's rotation gives all locations on Earth, except at the poles, a velocity to the east. The speed depends on the distance from the equator and decreases with the cosine of latitude. So the speed due to rotation is zero at the poles.

Example 1: Calculate the speed of a person at the equator due to Earth's rotation. This speed is one Earth's circumference per day, where the circumference is $2\pi R$, R being the radius of Earth (6,378 km). $2\pi R$ / 1 day = 464 m/s or 1670 km/hr. For comparison, cars on interstate highways typically travel 100 km/hr, while commercial jet aircraft usually cruise at a little over 1000 km/hr.

The speed from Earth's revolution about the Sun is Earth's orbital path length divided by one year. The direction of this velocity is along Earth's orbit. The direction of this velocity rotates slowly in the plane of the ecliptic, making one revolution per year. This velocity will appear to be changing direction continuously from a fixed location on Earth, because of Earth's rotation. It will be approximately east on Earth at solar midnight, while near the equator it will be approximately straight up at sunrise, and it will be close to west at solar noon. It is the tilt of Earth's rotational axis relative to the plane of the ecliptic that causes the eastward velocity from the Earth's rotation to swing daily from a maximum of 23.5° above the plane of the ecliptic to 23.5° below it. Figure 15-1 illustrate's the relative directions of these velocities.

A globe showing the tilt of Earth's axis provides a useful display for these velocities. A pointer, such as a pencil, laid on the table by the base of the globe, pointing away from the globe, can show the direction of the velocity from revolution. A second pointer held against

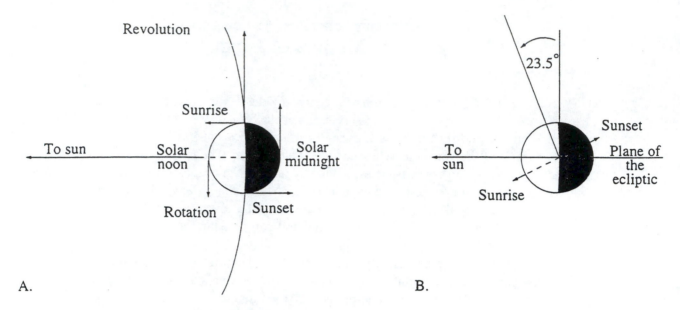

Figure 15-1. The relative directions of velocities from the motions Earth from rotation and revolution, A. as seen from above the ecliptic, and B. as seen from the plane of the ecliptic at the time of the summer solstice in the northern hemisphere.

and parallel to the equator, while turning the globe and second pointer together, will show in three-dimensions the relative directions of these velocities.

Some other motions, considered below, have higher speeds than those just considered, and collectively all of these motions are identified as **major motions** of Earth. Still other motions, also to be considered below, have much lower speeds and are considered to be **minor motions**. Major motions result from Earth's rotation and revolution, from the Solar System moving within the Galaxy, and from movements of the Galaxy.

The most significant motion inside the Milky Way galaxy is the Sun moving toward Vega at about 20 km/s, and moving with Vega in a nearly circular orbit through the Galaxy's disk and around the Galaxy's center. The orbit around the Galaxy occurs with a speed of about 240 km/s, and is in the direction toward the constellation Cygnus and away from Orion. The center of the galaxy is in the direction of the constellation Sagittarius. The circuit time, sometimes called a **galactic year**, is about 230 million years. These motions are detected by several methods, including the Doppler shifting of 21 cm radiation from hydrogen located through out the galaxy disk.

The Milky Way galaxy is also in orbit about the center of mass of a cluster of about 30 galaxies called **the Local Group**. The Local Group contains the Milky Way, Andromeda galaxy, the Large and Small Magellanic Clouds, all of which can be seen with the unaided eyes from Earth, and other galaxies. The Milky Way is moving with a velocity of about 80 km/s towards Andromeda galaxy. Additionally, the Local Group is moving collectively towards a local **supercluster**, the Virgo Cluster, which is in the direction towards the Hydra-Centaurus supercluster, beyond which lies a more massive supercluster, called the **Great Attractor**. Our motion in that direction was detected from variations in the **cosmic background radiation**, recorded by the **COBE** (Cosmic Background Explorer) satellite, and is believed to be due to a large mass concentration in the Great Attractor. The Local Group is moving with a speed of about 500 km/s towards the Great Attractor. A globe showing the locations of the constellations as they appear on the celestial sphere from Earth provides a useful model on which to show the relative directions of these velocities.

There are still other motions of Earth, with much smaller associated speeds than those just considered. These minor motions are included here because their effects are readily detected. One of these motions results from the Earth and Moon orbiting the center of mass of the Earth-Moon system. The period of this motion is about a month, and the path length is relatively short, so the speed is very low. Nevertheless, this motion helps create tides in Earth's oceans, it contributes to the precession of the Earth's axis, and also to slowing Earth's rate of rotation. The latter effect lengthens the day by about 2 milliseconds per century.

Another minor motion is associated with the precession of Earth's axis, which causes the celestial poles of Earth to move around a circle with an angular radius of 23.5°. The period of this motion is about 26,000 years. Because of this long period, the resulting speeds are extremely low, and are discussed here only because this motion causes the dates when the Sun enters different constellations of the zodiac to slowly change. The position of the Sun at the time of birth determines one's sign in the practice of **natal Sun-sign astrology**. Ancient astrology also considered the positions of the Moon and other then-known planets at the time of birth. Apparently this creates too many variations in horoscopes charts for today's tabloids. Even just using a simplified "Sun-only" system, modern tabloid astrology has not kept up with the slow precession of Earth's axis, as illustrated in the exercise on Star Wheels.

Acceleration of an object is defined as the rate its velocity changes. Acceleration, like velocity, is a vector. The units of acceleration are speed per time and are usually given as distance per second per second, such as m/s^2 or km/s^2. The acceleration due to gravity on the Earth's surface, usually called **g**, is 9.8 m/s^2 down. This means that an object dropped from rest will accelerate to a speed of 9.8 m/s after it has fallen for 1 second, ignoring any friction, such as with air. **g** provides a useful reference for other accelerations.

Dealing with gravity is an important part of every day life for people on Earth, and healthy adults are usually so accustomed to doing it that they don't often consciously think about it. But when a person loses consciousness for a moment, as when a boxer or other athlete is knocked out, the effects of gravity on that person, and their temporary inability to cope with it, are immediately apparent. Gravity continually pulls us down. When we think about it, we can feel the forces we use to stop ourselves from being accelerated downward and into whatever is below us.

Accelerations from the various motions of Earth are all very small compared to **g**, and so may be expected to go mostly unnoticed. The small values of these accelerations may also be surprising because of the high associated speeds described above. But small accelerations accumulated for long periods of time may result in high speeds.

Example 2: Find the velocity achieved by an object that is accelerated by **g** for six months.
$$v = (9.8 \text{ m/s })\cdot(1.6 \cdot 10^6 \text{ s}) = 1.6 \cdot 10^8 \text{ m/s}.$$
Note that this value is just over half the speed of light. **g** is a large acceleration.

When an object travels around a circle at constant speed, the velocity changes direction continuously. This change is defined as **centripetal acceleration**, and this acceleration results from the force that produces the circular motion. Many amusement park rides take advantage of centripetal acceleration. For example, one may ride inside a large rotating barrel. While spinning, the bottom of the barrel can be removed. The ride is exciting because it looks like one will fall, but no one falls because they are being pressed out against the barrel sides. The centripetal acceleration is toward the center of the barrel, and has the magnitude, $a_c = v^2 / r$, where v is the velocity and r is the radius of the circular motion. For orbital motion, the velocity is the orbit length divided by the orbit period T, or $v = (2\pi r) / T$. Inserting this value of v into the equation for a_c, we have, $a_c = (4\pi^2 r) / T^2$. This formula will be used below. Note that the units of a_c are just those of acceleration, m/s^2.

When a satellite orbits a massive body, the gravitational force of attraction from the massive body accelerates the satellite, creating the satellite's centripetal acceleration. The satellite is said to be weightless, or in free fall, as it is "falling around" the central massive body. The environment inside an orbiting space shuttle is described as being **micro-g**, indicating that parts of the shuttle closer to Earth are attracted to Earth by slightly more force than parts further away. But the structure of the shuttle keeps all of its parts moving together as one body. These different forces of gravity acting on different parts of a body are classified as **tidal** forces, and their differences become larger as the orbiting and orbited objects become larger, more massive, and closer together. A prominent theory on the origin of planetary rings is that they result when tidal forces break apart satellites that stray too close to a planet. Comets have been observed to break into pieces near planets, verifying such effects.

Earth experiences centripetal acceleration from its revolution about the Sun, and also from co-orbiting with the Moon. If it were not for the Moon, the center of mass of the Earth would move in an elliptical orbit around the Sun. But the Moon's presence causes the Earth and Moon to co-orbit their common center of mass, as that center of mass orbits the Sun. Consider the motion of a hypothetical double planet consisting of two planets with equal mass. The center of mass of this pair would be halfway between them, and that center of mass would travel in an ellipse about their star.

Example 3: Find the distance from the center of Earth to the center of mass of the Earth-Moon system.

The location of the center of mass away from the center of Earth is $D_c = D_M M_M/(M_M + M_E)$, where D_M is the distance from the center of Earth to the center of the Moon, and M_M and M_E are the masses of the Moon and Earth, respectively. The usual values give, $D_c = 4{,}674$ km, which, since the radius of Earth is 6,378 km, is still inside Earth.

Example 4: Determine the velocity and acceleration that results from Earth co-orbiting the Moon, at the place on Earth's surface closest to the Moon, and also at the place furthest from the Moon.

On the side of Earth closest to the Moon, the distance from the Earth-Moon center of mass is 6378 - 4674 km = 1704 km, while on the side farthest from the Moon the distance is 6378 + 4674 km = 11,052 km. The centripetal accelerations at these locations are, $(2\pi)^2$ x 1704 km/(1 month)^2 = 0.010 mm/s^2, and $(2\pi)^2$ x 11,052 km/(1 month)^2 = 2.3 x 10^{-8} km/s^2 = 0.023 mm/s^2 respectively.

The effects of tidal forces from Earth on the Moon have caused the Moon's rotation and revolution about Earth to become synchronized, so the same side of the Moon faces Earth all of the time. This same process is slowing Earth's rotation. If it could proceed for long enough, it would ultimately synchronize the rotation of Earth and revolution of the Moon, causing the lunar month to equal the solar day, which would then be equal to about 47 of our present solar days. The time required for this to happen is estimated to be over 40 billion years, longer than the Sun is expected to last. However, Pluto and its moon, Charon, have already achieved this stable configuration.

The magnitude of the centripetal acceleration from the Earth-Moon system's revolution about the Sun is, $(2\pi)^2$ x 1.50 x 10^8 km/(1 year)^2 = 5.9 x 10^{-6} km/s^2 = 5.9 mm/s^2. Notice that this is small compared to g, 9.8 m/s^2. So one should not expect to feel this acceleration, although it does have observable effects. For example, it contributes to tides in Earth's oceans. The direction of this acceleration vector is towards the Sun.

Procedures

Apparatus

calculator, colored pencils, globe of Earth showing Earth's tilt, globe of celestial sphere showing the constellations, stick-on notepads, and scissors.

A. Comparing Speeds

1. Calculate the speed associated with the Earth's revolution in km/hr and m/s. Assume Earth's orbit is a circle with a radius 1 AU, where 1 AU = 150. x 10^6 km.

2. Verify the value given in the discussion above for the speed of the Sun moving about the center of the Milky Way, assuming the Galaxy's center is 35 kly distant, and the galactic year is 230 million years. Note that one kly is one thousand light-years. This can be converted to meters by multiplying one thousand times the speed of light in meters per second times the number of seconds in one year.

3. Make a table listing the speeds associated with all motions of Earth, those you calculated as well as those given in the text above, arranged from highest to lowest.

4. Draw a circle of about 10 cm radius to represent the Earth's orbit about the Sun. Use different colors to show the paths of the Earth and Moon, and greatly exaggerate their separation, to say 10% of the distance to the Sun. Show the paths followed by the Earth and Moon, as they execute sinuous trajectories, making 12 cycles (one per month) in one revolution.

5. Determine at what time(s) of day and what days of the year your velocity from Earth's rotation will be in the same direction as your velocity from the Earth's revolution. Hint: use pencil pointers to show these two velocities on a globe that shows Earth's tilt, and consider only special days, such as the days of the vernal and autumnal equinoxes and the summer and winter solstices.

6. Construct stick-on note pointers as shown in Figure 15-2. Then place these pointers on a globe of the celestial sphere showing the constellations, so the pointers show the directions of the velocity of the Solar System within the Galaxy, and the velocity of the Galaxy through space.

B. Comparing Accelerations

1. Verify the value reported above by calculating the value of the centripetal acceleration for the Earth in its revolution about the Sun.

2. Calculate the centripetal acceleration produced by the Solar System orbiting the center of the Milky Way galaxy.

3. Make a table of accelerations associated with all of the major motions of Earth, arranged from highest to lowest.

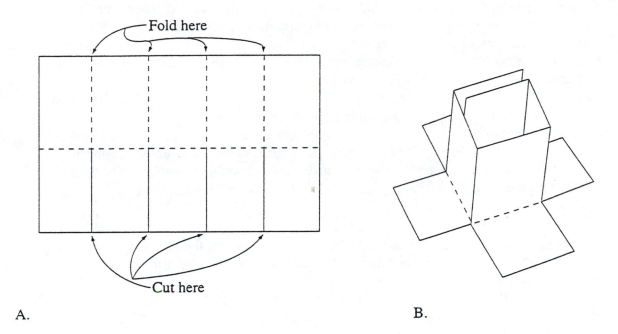

A. B.

Figure 15-2. Construct stick-on note pointers by making four equally spaced cuts through the adhesive side of a stick-on note, as shown in A., and fold that note into a box, as shown in B., with the adhesive side in. Then, fold the flaps created by the cuts out, as shown, and stick the pointer created onto the celestial sphere, as described in the text.

Kepler's Laws of Planetary Motion I

In struggling to understand the motions of the planets in the sky, Johannes Kepler empirically worked out three basic laws, now known as Kepler's Laws of Planetary Motion. The first law (Law of Ellipses) states that the orbit of a planet is an ellipse, with the Sun at one focus. The second law (Law of Equal Areas) states that a line drawn from a planet to the Sun sweeps out equal areas in equal amounts of time. The third law (Harmonic Law) states that the cube of the semi-major axis is proportional to the square of the sidereal period. Kepler developed these laws to apply to a single planet orbiting the Sun. However, Kepler's Laws provide good approximations of the motions of many orbiting bodies. This laboratory exercise concerns the first two laws.

The first law states that an orbiting body travels in an ellipse. An ellipse is a two-dimensional shape for which the sum of the lengths of a line drawn from one focus to a point on the ellipse and another line drawn from the other focus to that same point remains constant for every point along the ellipse. Usually, when describing an ellipse, one talks about its eccentricity. Referring to Figure 16-1, the eccentricity is defined as

$$e = \frac{c}{a} = \frac{a-b}{a}, \tag{1}$$

where c is the distance from the center of the orbit to a focus, a is the semi-major axis, and b is the semi-minor axis. From this definition, the closest point of a planet from the Sun (perihelion distance) is

$$R_p = a - c = a - ae = a(1-e). \tag{2}$$

Similarly, the aphelion (farthest distance from the Sun) is

$$R_a = a + c = a + ae = a(1+e). \tag{3}$$

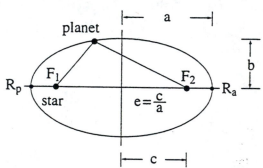

Figure 16-1: An ellipse. a is the semi-major axis, b is the semi-minor axis, c is the distance from the center to a focus, e is the eccentricity, R_p is the peri-distance, and R_a is the apo-distance.

Example 1: To test the Law of Ellipses, consider the Moon orbiting Earth. The Moon's orbital eccentricity is given as 0.055, and its semi-major axis is equal to 384,400 km. On 12 August 1994, the Moon is at perigee at a distance of 369,453 km. On 27 August 1994, the Moon is at apogee at a distance of 404,332 km. Check these values using the above formulae.

$$R_p = a(1-e) = (384400 \cdot km)(1 - 0.0550) = 363000 \cdot km$$
$$R_a = a(1+e) = (384400 \cdot km)(1 + 0.0550) = 406000 \cdot km$$

These predicted values are within 2 percent of the actual values.

The Law of Equal Areas deals with the speed of the orbiting object. A line connecting the Sun to a planet sweeps out equal areas in equal periods of time. Consider Figure 16-2. A line drawn between the Sun and the planet at point 1 moves around with the planet for, say, one month to point 2. The line is said to sweep out the pie-slice area A. During a similar one month period, the line sweeps out the area B as the planet travels from point 3 to point 4. Kepler's second law states that these areas are equal. In order for this to be true, the distance traveled by the planet when farther from the Sun (point 3 to point 4) must be shorter than the distance traveled by the planet when closer to the Sun (point 1 to point 2). Since the amount of time is the same, the planet must be traveling faster when closer to the Sun than when farther away.

The speed of the planet at any point in its orbit may be determined from

$$v = \sqrt{\left(\frac{4\pi^2 \cdot a^3}{P^2}\right)\left(\frac{2}{r} - \frac{1}{a}\right)}, \qquad (4)$$

where a is the semi-major axis of the planet's orbit, P is the sidereal period of revolution, and r is the current distance from the Sun. As with the Law of Ellipses, Kepler's Second Law can be applied to any body orbiting another.

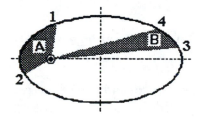

Figure 16-2: The planet sweeps out the equal areas, A and B, in equal amounts of time. The planet orbits from point 1 to point 2 in the same amount of time it takes it to orbit from point 3 to point 4.

Example 2: The Moon's perigee and apogee distances were determined in Example 1. Using these distances and Equation 4, the Moon's maximum and minimum speeds can be computed. Thus,

$$v_{max} = \sqrt{\left(\frac{4\pi^2 \cdot (384400 \cdot km)^3}{(2.36 \cdot 10^6 \cdot sec)^2}\right)\left(\frac{2}{363258 \cdot km} - \frac{1}{384400 \cdot km}\right)} = 1.08 \cdot \frac{km}{sec}$$

$$v_{min} = \sqrt{\left(\frac{4\pi^2 \cdot (384400 \cdot km)^3}{(2.36 \cdot 10^6 \cdot sec)^2}\right)\left(\frac{2}{405542 \cdot km} - \frac{1}{384400 \cdot km}\right)} = 0.969 \cdot \frac{km}{sec}.$$

Procedures

Apparatus

none.

Table 1: Some moons of Jupiter.

moon	semi-major axis (km)	eccentricity	sidereal period (days)	diameter (km)
Amalthea	180,500	$3.00*10^{-3}$	0.498	270*
Io	422,000	0.000	1.769	3,630
Europa	671,000	0.000	3.551	3,138
Ganymede	1,070,000	$2.00*10^{-3}$	7.155	5,262
Callisto	1,883,000	$8.00*10^{-3}$	16.689	4,800
Himalia	11,480,000	0.158	250.57	180*
Sinope	23,700,000	0.275	758.0	40*

* Some moons are not spherical, the diameter given is for the longest axis.

1. Using Table 1 and Equations 2 and 3, construct a table of perijovian and apojovian distances for these seven moons of Jupiter.

2. Of the above seven moons, which have circular orbits? Which have the most eccentric orbits?

3. Use the data constructed in part 1 to construct a table of minimum and maximum speeds for the seven jovian moons listed in Table 1. Express these speeds in kilometers per second.

4. On average, the Moon subtends an angle of 0.5° (or 30') in our sky. Add to the table created above two columns containing the angular size of each moon at perijovian and apojovian as seen from the cloud-tops of Jupiter. Figure 16-3 shows a right triangle formed by a line from the observer's position in the cloud-tops of Jupiter to one side of the moon, across the diameter of the moon, then back to the observer. The subtended angle is the angular size of the moon as seen from the jovian cloud-tops. The radius of Jupiter is 71,400 km. Record your answers in minutes-of-arc.

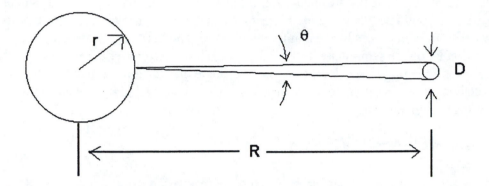

Figure 16-3: From the surface of a planet of radius r, the angular size θ of a moon of diameter D at a distance of R from the center of the planet is given by

$$\theta = \arctan\left(\frac{D}{R-r}\right).$$

Kepler's Laws of Planetary Motion II

In struggling to understand the motions of the planets in the sky, Johannes Kepler empirically worked out three basic laws, now known as Kepler's Laws of Planetary Motion. The first law (Law of Ellipses) states that the orbit of a planet is an ellipse, with the Sun at one focus. The second law (Law of Equal Areas) states that a line drawn from a planet to the Sun sweeps out equal areas in equal amounts of time. The third law (Harmonic Law) states that the cube of the semi-major axis is proportional to the square of the orbital period. Although Kepler developed these laws to apply to a single planet orbiting the Sun, they provide excellent approximations to many orbital scenarios. This laboratory exercise will explore the Jovian System with the third law.

The Harmonic Law may be stated as
$$a^3 = kP^2, \qquad (1)$$
where a is the semi-major axis of the orbit, P is the sidereal period of revolution, and k is a constant of proportionality called the Kepler constant. This constant depends only on the body being orbited and not the orbiting body. So, once the Kepler constant is known for the central body, all objects in orbit about it will use that constant. One can solve for the Kepler constant if one knows the orbital characteristics of at least one body.

Example 1: The Kepler constant for the Sun can be determined from the orbital characteristics of Earth.

Earth orbits the Sun at an average distance of 1 astronomical unit (AU) in 1 year (y). So, the Kepler constant for the Sun is

$$k = \frac{a^3}{P^2} = \frac{(1 \cdot AU)^3}{(1 \cdot y)^2} = 1 \cdot \frac{AU^3}{y^2}.$$

Example 2: Jupiter can be observed to take about 11.86 years to complete one orbit of the Sun. What is the average distance of Jupiter from the Sun?

The average distance of a planet from the Sun is the same as its semi-major axis, so we apply the Harmonic Law to this problem. Thus,

$$a = \sqrt[3]{k \cdot P^2} = \sqrt[3]{\left(1 \cdot \frac{AU^3}{y^2}\right)(11.86 \cdot y)^2} = 5.20 \cdot AU.$$

On average, Jupiter orbits the Sun at a distance of 5.20 times the average distance of Earth from the Sun.

Example 3: The Kepler constant of Earth may be found using the Moon's orbit.

The Moon orbits at an average distance of 384,400 kilometers (km) in 27.3 days (d). Thus, the Kepler constant of Earth is

$$k = \frac{a^3}{P^2} = \frac{(384400 \cdot km)^3}{(273 \cdot d)^2} = 7.62 \cdot 10^{13} \cdot \frac{km^3}{d^2}.$$

Now, since a communications satellite also orbits Earth, we can find the a or P for a satellite using this value of k. A geosynchronous satellite revolves at the same rate that Earth rotates (i.e., one revolution per day). The distance of this satellite from the center of Earth is

$$a = \sqrt[3]{kP^2} = \sqrt[3]{\left(7.62 \cdot 10^{13} \cdot \frac{km^3}{d^2}\right)(1.00 \cdot d)^2} = 42400 \cdot km.$$

Hence, a body in orbit at a distance of 42,400 km from the center of Earth will revolve once a day. Since Earth has a radius of 6378 km, the satellite is roughly 36,000 km above the surface. This is the altitude of most communications satellites.

The Harmonic Law was later derived by Sir Isaac Newton from the basic principles of rotary motion and gravitation. The gravitational force, F, between two masses is given by

$$F = G \cdot \frac{Mm}{r^2}, \tag{2}$$

where G is the Universal Gravitational constant ($6.672 \cdot 10^{-11} \cdot \frac{N \cdot m^2}{kg^2}$),

M and m are the masses of the two bodies, and r is the distance between the centers of the two bodies. The unit of force in the MKS system is the newton (N) and is equivalent to the force needed to accelerate a 1 kg mass at 1 m/s^2.

Example 4: An 80.0 kg being standing on Earth's surface is attracted toward the center of Earth by the force of gravity. Earth has a mass of $5.98 \cdot 10^{24}$ kg and a radius of 6378000 m.
From Equation 2, the force of gravity acting on the being is

$$F = \left(6.672 \cdot 10^{-11} \cdot \frac{N \cdot m^2}{kg^2}\right) \frac{(5.98 \cdot 10^{24} \cdot kg)(80.0 \cdot kg)}{(6378000 \cdot m)^2} = 785 \cdot N.$$

Thus, the weight of the being is 785 newtons on Earth's surface. On the Moon, whose mass and radius are $7.35 \cdot 10^{22}$ kg and 1738 km, respectively, an 80.0 kg being would weigh only 130. N.

Newton's second law of motion states that the force on a mass, m, undergoing acceleration, a, is

$$F = ma. \tag{3}$$

Since Equations 2 and 3 both give the force acting on the orbiting body, one can solve for the acceleration due to gravity (g). Thus,

$$g = \frac{GM}{r^2}. \tag{4}$$

Notice that the mass of the orbiting body has been eliminated from the formula.

Example 5: The acceleration due to the gravity of Earth at its surface (g) is given by Equation 4,

$$g = \frac{GM}{r^2} = \frac{G \cdot (5.98 \cdot 10^{24} \cdot kg)}{(6378000 \cdot m)^2} = 9.81 \cdot \frac{m}{s^2}.$$

Example 6: Another way to determine the weight of a being is to use Newton's second law (Equation 3). If we again consider an 80.0 kg being on the surface of Earth, its weight would be

$$F = mg = (80.0 \cdot kg)\left(9.81 \cdot \frac{m}{s^2}\right) = 785 \cdot N.$$

This is the same weight computed in Example 4.

The centripetal acceleration, a, on an orbiting body is given by

$$a = \frac{v^2}{r}, \tag{5}$$

where v is the average tangential velocity of the orbiting body, and r is the average distance from the body to the center of the orbit. Since the planet stays in orbit, the centripetal acceleration must equal the acceleration due to gravity. Setting the right-hand side of Equation 4 to the right-hand side of Equation 5, one has

$$v^2 = \frac{GM}{r}. \tag{6}$$

This average tangential velocity of an orbiting mass is given by the length of the orbit divided by the amount of time it takes to go around the orbit (its period of revolution), or

$$v = \frac{2\pi \cdot r}{P}, \tag{7}$$

where P is the sidereal period of the orbiting mass and r is the average radius of the orbit.

Now, substituting Equation 7 into Equation 6 and rearranging, gives

$$r^3 = \left(\frac{GM}{4\pi^2}\right) \cdot P^2. \tag{8}$$

This is Newton's restatement of Kepler's third law of planetary motion. The Kepler constant (k) is, thus,

$$k = \frac{GM}{4\pi^2}. \tag{9}$$

Example 6: Equation 9 can be used to compute the mass of the central body if the Kepler constant is known. The Kepler constant of Earth ($7.62 \cdot 10^{13}$ km^3/d^2) was computed in Example 3. This value must be converted to base MKS units before it can be used in Equation 9.

Thus,

$$k = 7.62 \cdot 10^{13} \cdot \frac{km^3}{d^2} * \left(\frac{1000 \cdot m}{1 \cdot km}\right)^3 * \left(\frac{1 \cdot d}{86400 \cdot s}\right)^2 = 1.02 \cdot 10^{13} \cdot \frac{m^3}{s^2}.$$

Solving Equation 9 for the mass of the central body (M),

$$M = \frac{4\pi^2 \cdot k}{G} \tag{10}$$

Now, the mass of Earth is

$$M = \frac{4\pi^2 \cdot \left(1.02 \cdot 10^{13} \cdot \frac{m^3}{s^2}\right)}{\left(6.672 \cdot 10^{-11} \cdot \frac{N \cdot m^2}{kg^2}\right)} = 6.04 \cdot 10^{24} \cdot kg$$

If this method is used with more accurate data, the mass becomes $5.98 \cdot 10^{24}$ kg, which is only 1 percent from the rough calculation performed above.

Most astronomy magazines include graphs representing the positions of the largest moons of Jupiter and Saturn during the month. These 2-D plots of the orbits appear as sine waves. Figure 17-1 is just such a plot of the Galilean satellites, Io, Europa, Ganymede, and Callisto. The vertical axis of the graph represents time, with each horizontal line marking a day. The horizontal axis of the graph represents distance from the center of Jupiter. The two vertical lines in the center of the graph represent the eastern and western limbs of Jupiter. The distance between the two vertical lines gives the diameter of Jupiter on the scale of the graph.

The period of a moon can be determined by counting the number of days between consecutive peaks on the wave (its wavelength). The amplitude of the wave, measured from halfway between the two vertical lines to the peak of a moon's wave, gives the semi-major axis of the moon.

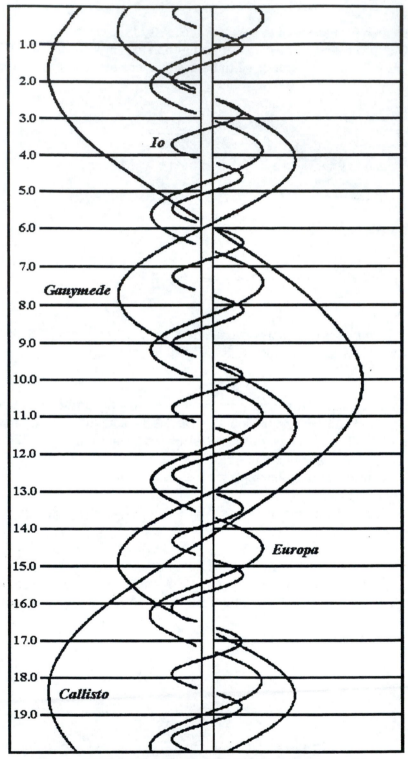

Figure 17-1: A plot showing the relative positions of the
Galilean Moons of Jupiter, as seen from Earth on the days indicated.

Procedures

Apparatus
current 2-D graphical representation of orbital data for the Galilean satellites, millimeter rulers.

1. Determine the scale along the vertical axis of the 2-D plot by measuring the distance between consecutive day marks in millimeters. This will allow you to determine the period of a moon to the hour. Such accuracy is necessary.

2. Determine the scale along the horizontal axis of the 2-D plot. Here, we are going to use the radius of Jupiter as our unit of distance. Measure the distance between the two vertical lines in the center (which is to scale with the diameter of Jupiter), and divide this by two. Whenever you measure the amplitude of a wave, this scale value will convert that number to jovian radii (j.r.).

3. Make a table containing the periods (in days) and semi-major axes (in jovian radii) of the four Galilean moons. To measure the period, simply measure the distance (in mm) between two consecutive peaks along the vertical axis. Multiply this measurement by the vertical scale. To measure the semi-major axis distance for a moon, measure from the peak of the wave to the center of the double lines, then multiply by the horizontal scale factor.

4. Using the data collected in part 3 and Equation 1, calculate the Kepler constant of Jupiter, k_J. What are the units of this constant of proportionality? Include these units in the column header in your table. The Kepler constant should be the same for each moon, since they all orbit Jupiter, but printing and measuring errors may cause slight differences.

5. Compute the arithmetic mean of these values for the jovian Kepler constant, k_J. This mean value should be used for all bodies orbiting Jupiter.

6. Sinope has a period of 758 days. Find the semi-major axis of this moon's orbit in both jovian radii and kilometers using k_J and Equation 1. (Note: 1 j.r. = 71,400 km.) Compare the semi-major axis of the orbit of Sinope, calculated above, to the published value of $2.37 \cdot 10^7$ km (i.e., compute the percentage error of the calculated semi-major axis to the published value).

7. Using k_J and Equation 10, determine the mass of Jupiter. Hint: you must convert k_J to standard MKS units first. Why is this so? The published value is $1.90 \cdot 10^{27}$ kg. Compare the calculated value to the published value by computing the percentage error.

Orbiting Earth

   Astronomy is progressing at an astonishing rate due in part to a variety of robot spacecraft and orbiting observatories that provide images of celestial objects in the x-ray, ultraviolet, infrared, and microwave portions of the electromagnetic spectrum. Even in the visible region of the spectrum, images from telescopes in space are used in making significant discoveries at an unprecedented rate. It is clear that astronomy, as it exists now, depends on the successful operation of robot spacecraft and observatories orbiting Earth. Thus, the geometry, mechanics, and technology of space transportation has become an important part of astronomy.

   Isaac Newton (1642-1727) was aware that it is possible to orbit Earth if sufficient speed can be achieved. Just as an object dropped near Earth is accelerated toward Earth by gravity, so too an object on orbit near Earth is accelerated toward Earth. One can think of a body on a circular orbit as a body that is falling toward Earth while it travels above the surface at such a rate that it is in effect falling around Earth, but never getting closer to it. The speed required to achieve circular orbit is,

$$V = \left(G \cdot \frac{M}{R}\right)^{\frac{1}{2}} \tag{1}$$

where G is the universal gravitational constant, M is the mass of Earth, and R is the radius of the orbit. Rockets capable of achieving orbital speeds were first developed during the second half of the 20th century, making it possible to conduct astronomical observations from above Earth's atmosphere.

   Example 1: Calculate the speed required to orbit Earth just above the atmosphere. The value of the universal gravitational constant is $6.67 \cdot 10^{-11}$ Nm^2/kg^2, the mass and radius of Earth are, $5.98 \cdot 10^{24}$ kg and $6.37 \cdot 10^6$ m. The radius of the orbit required is about 130 km plus the radius of Earth. To three significant figures, this is $6.50 \cdot 10^6$ m.
   The above formula and values yield, 7830 m/sec or 28,200 km/hr.

   Example 2: Calculate the Moon's speed in its orbit about Earth. Assume the Moon's orbit is circular with a radius of 384,000 km.
   The above formula and values yield, 1020 m/sec or 3670 km/hr.

   For a machine to orbit Earth, it is necessary for it to first get above the dense part of the atmosphere, then acquire "orbital" velocity horizontally. The primary reason rockets take off vertically is to use their powerful engines to lift the crafts and carry them through the dense, lower part of the atmosphere at low speed. Traveling at high speeds in the atmosphere would require strong, and therefore heavy, airframes and also would be accompanied by the continual loss of significant energy to friction with the atmosphere. Thus, it is possible to maintain orbital speeds only above most of the atmosphere.

Satellites on orbit about Earth obey Kepler's laws (see Exercises 16 and 17, Kepler's Laws of Planetary Motion I and II) and move in elliptical orbits at prescribed speeds with Earth's center of mass at one focus. A satellite's speed, direction, and distance from Earth at the time it is released or its rocket engines are turned off, will determine the size, shape, and orientation of the orbit. The main force acting on a satellite orbiting above the atmosphere is gravity, and gravity will cause the satellite to pass through the point where it was released (or its engine shut off) on each circuit. Satellites on orbit are often referred to as being in **free fall**, and astronauts and other object accompanying them experience **weightlessness**.

The closest point of an orbit to Earth is its **perigee** and the farthest point is its **apogee**. The perigee and apogee are on opposite sides of the orbit's major axis. The orbit is fixed relative to Earth's center of mass in space, independent of Earth's rotation. The plane of the orbit must always include Earth's center of mass. If an orbit is circular, the perigee and apogee are at the same distance from Earth's center of mass, and the satellite's speed is constant. Orbits just above the atmosphere are called **low orbits**, **low Earth orbits**, or **LEO's**.

Inclination is the angle an orbit makes with the plane of the equator, where this angle is measured counter-clockwise from the equator where the satellite crosses the equator heading north. Thus, it is possible to have an orbit with inclination greater than 90 degrees. An orbit directly over Earth's equator has an inclination of zero degrees, and is called an **equatorial orbit**. An orbit that passes over both poles has an inclination of 90 degrees, and is called a **polar orbit**. A satellite on polar orbit above the terminator (the line dividing the sunlit hemisphere from the dark hemisphere of a planet or moon) would be in sunlight all of the time. This is the only orbit which can provide continuous solar power, and is one type of **sun-synchronized orbit**.

Another type of sun-synchronized orbit is a polar orbit in which a satellite crosses the equator at the same local solar time on each orbit. Landsats used this orbit so the images of Earth they provided were obtained with essentially the same angle of illumination on each orbit, and therefore could be directly compared. The images from these satellites have been very useful in studying the planet Earth. Satellites that require large amounts of electric power when they operate out of the sunlight must carry rechargeable batteries or nuclear (i.e., radioactive) generators.

The **ground trace** of a satellite is the track it makes over the surface of Earth. The highest northern and southern latitudes reached by a satellite's ground trace is equal to the satellite's orbital inclination (or 180 degrees minus that angle, if it is greater than 90 degrees). Ground traces often make unusual patterns when drawn on a flat map of Earth, but are revealed to be "great circles" when shown on a globe. The great circles are where the plane of the orbit intercepts the surface of Earth. The ground trace for a satellite in polar orbit when it comes down over the north pole, heading due south, is a line tilted toward the west because of Earth's rotation. The same is true as the satellite heads north from the south pole.

The **period** of an orbit is the time it takes the satellite to complete one circuit, and can be calculated for circular orbits with the equation,

$$P = \left(\frac{4 \cdot \pi^2 \cdot R^3}{G \cdot M} \right)^{\frac{1}{2}} \tag{2}$$

if the radius, R, is known. Note that the shortest period occurs with the lowest orbit.

Example 3: Find the minimum period for an orbiting satellite.
The minimum period would occur for an orbit just above Earth's atmosphere. Using the values for G, M, and R given in Example 1 above, with Equation 2 yields, 5210 seconds, 86.9 minutes or 1.45 hours.

This equation may be manipulated to give the radius of a circular orbit for a particular period, P, as,

$$R = \left(\frac{P^2 \cdot G \cdot M}{4 \cdot \pi^2} \right)^{\frac{1}{3}} . \tag{3}$$

Example 4: Find the radius of an orbit that would cause a satellite to orbit Earth with a period of one sidereal day, 23 hours and 56 minutes.
Equation 3 and this period yields a radius of 42,200 km. A satellite on orbit at that radius would be 35,800 km above Earth's surface.

A satellite on an equatorial orbit at 35,800 km above Earth traveling east, is in a **geostationary** or **geosynchronous orbit**, often called **geosynch**. Many communications and weather satellites are on geosynch. From vantage points on Earth, these satellites appear fixed in the sky. "Satellite" dishes aimed at these points may receive microwave signals from the satellites 24 hours a day. From the satellite's viewpoint, its transmitter can send signals to, and its cameras can see, nearly one-third of Earth's surface, although it cannot effectively reach high (north or south) latitudes. The ground trace for such an orbit is a spot on the equator. Geosynch is an example of a **high orbit**.

Satellites on geosynch and other orbits are subjected to small gravitational forces, perturbations, from the Moon, Sun, and the oblate shape of Earth, which produce small accelerations that accumulate over time and cause the satellite to rotate and also to move out of their desired positions. "Station keeping" is required to keep satellites oriented properly, so cameras, telescopes and radio antennae are pointed in the correct directions, and also to maintain the velocity required by the orbit. Electrically powered gyroscopes rotating around different axes on a satellite can be sped up or slowed down to maintain rotational control of the satellite, and small "thruster" rockets can be momentarily fired to correct the satellite's velocity. The useful lives of many satellites on high orbits end when their gyroscopes fail, or their thrusters run out of fuel. The space shuttle can provide service calls to satellites on low orbits, but currently there is no capacity for service calls to high orbits.

An ideal location for a space port would be the top of the highest mountain on Earth's equator. This would allow satellites to achieve orbit with the minimum expenditure of rocket fuel. Launching eastward takes advantage of Earth's rotation, which provides a velocity of 1670 km/hr eastward at the equator. See Exercise 15, Motions of Earth. The Kennedy Space Center, at Cape Canaveral, Florida, is the primary launch facility of the United States space program. The location was chosen to take advantage of year-round warm weather, the availability of then undeveloped land, of being relatively close to Earth's equator in then "secure" US territory, and also to be on the eastern seaboard. Being on the eastern seaboard places rockets that fail early during their launches over the Atlantic ocean, where there are no cities to be damaged by the falling debris. The facilities used by the European Space Agency in French Guiana, are well situated by being near the equator, while the Russian Cosmodrome at Biakonur is not. Japan, China and India also have launched Earth orbiting satellites from their own territories.

To achieve orbit, a rocket must change its velocity from whatever it is to the orbital speed described above. Firing a rocket engine for a measured time while the rocket is in flight will produce a change in velocity, often referred to as ΔV (pronounced delta vee). Larger ΔV's require more fuel. All of the fuel consumed by a rocket in flight must be lifted by the rocket engines at take-off.

Example 5: Calculate the ΔV required to put a satellite in LEO by launching eastward, southward, and westward from the equator.
Orbital speed for LEO, calculated in example 1, is 28,200 km/hr, but when launching eastward from the equator the ΔV needed is lessened by the rotational speed of Earth, 1670 km/hr. Thus, 28,200 - 1,670 = 26,530 km/hr is the required speed for an eastward launch. Launching southward, the ΔV needed is just 28,200 km/hr. When launching westward, the ΔV needed is the 28,200 + 1670 = 29,870 km/hr.

Acquiring another orbit in the same orbital plane (i.e., a higher or lower orbit) requires changing speeds (ΔV's). While on orbit this is accomplished by firing the rocket engines. To move to a higher orbit, the rocket engines are fired so as to speed up the spacecraft, which causes the spacecraft to move higher. This, in turn, causes it to slow down, so it ends up higher but traveling slower than before the rocket engines were fired. To move to a lower orbit, the rocket engines are fired to slow the spacecraft down, which causes it to fall to a lower orbit. This, in turn, causes it to speed up. Orbiting spacecraft could not carry enough fuel to slow down sufficiently to land on Earth, if they could not use friction from traveling through Earth's atmosphere. Since the Moon has no atmosphere, the Apollo spacecraft had to carry substantial fuel from Earth to use to slow down sufficiently to land on the Moon. Changing the inclination of the orbit of an Earth orbiting spacecraft requires so much rocket fuel that it is usually not possible to change the inclination significantly.

Rocket scientists and engineers are continuing to develop engines that provide increased values of ΔV per kilogram of consumed fuel and oxidizer. Scientific and technical equipment are being made smaller and of lower mass, which reduces the mass that needs to be lifted. This has made it possible to achieved orbit by launching in any direction, but launching eastward still allows lower cost orbits, and so are usually the orbits of choice.

Satellites are often observed passing overhead just after sunset or before sunrise, when the sky is rather dark, but the Sun is not too far below the horizon. At this time, the satellites are illuminated by sunlight, while the observer below is in darkness. If the directions are known to the observer, it is fairly easy to distinguish polar from inclined and equatorial orbits. The National Aeronautics and Space Administration (NASA) maintains a site on the Internet which provides up-to-date information on their launches from Kennedy Space Center, and many people observe Space Shuttle launches. It is not possible to get very close to the launch sites, so telescopes are useful for viewing and photographing the launches.

Earth turns under the orbit, which alters the orbital period of a satellite as measured from a fixed point on the ground. In the 87 minutes calculated for the minimum orbital period, Earth turns about 22 degrees, so a satellite on a low Earth orbit of zero degrees inclination will require longer than 87 minutes to pass overhead again. The period of the orbit calculated above will not be the period of the orbit observed on Earth, because of the Earth's motion. The synodic period, P', can be calculated from,

$$\frac{1}{P'} = \frac{1}{P} - \frac{1}{E}, \tag{4}$$

where P is the sidereal period and E is the period of Earth's rotation. This equation can be rearranged to give,

$$P' = \frac{(E \cdot P)}{(E - P)} \tag{5}$$

Example 6: Calculate the period that would be observed by a fixed observer for the orbit just considered.

The orbital time can be calculated from the Equation 5,

$$P' = \frac{(1440 \cdot \min)(87 \cdot \min)}{(1440 \cdot \min - 87 \cdot \min)} = 93 \cdot \min$$

Procedures

Apparatus

globe of Earth, rolls of removable tape, drawing compass, millimeter scale, meter stick.

A. Drawing Orbits to Scale

1. Use a drawing compass and draw as accurately as possible a circle to represent Earth at the scale of 1.00 cm = 1000 km. Then, using the same center and scale, draw a circle to represent the orbit of a satellite 130 km above Earth.

2. Use a drawing compass to draw another circle to represent Earth at the scale of 1.00 cm = 5000 km. Then, using the same center and scale, draw a circle to represent the orbit of a satellite at geosynch.

3. Draw free hand, as accurately as possible, a circle to represent Earth at the scale of 1.00 cm = 50,000 km. Do this by placing two small marks a distance apart to represent the diameter of Earth, and then sketch a circle of that diameter. Then, using the center of that circle, and the same scale, draw a circle with a drawing compass to represent the orbit of the Moon.

4. On the drawing showing the orbit of the Moon around Earth, add to scale as accurately as possible a low Earth orbit and a geostationary orbit.

5. Comment on the difficulty of accurately showing these three orbits on the same drawing.

B. Showing Orbits on a Globe

1. Measure and record the diameter of your globe in cm. Compute the scale factor for your globe (a ratio which will convert cm to km for your globe). Then calculate the number of centimeters from the globe's surface a satellite would be on geosynch. Record that number in your report.

Position a meter stick perpendicular to the surface with one end on the equator of your globe, and place your head against the meter stick so one eye is at the equivalent of the geosynch distance from the globe. Look at the globe from that point, and make a list of which continents can be seen as the globe is rotated. Can all continents be seen? Which can not? What are the highest and lowest latitudes that can be seen?

What longitude (on the equator) would be the best location for a communications satellite on geosynch to serve both USA and Europe? The USA and Japan? Japan and Australia? Is it possible for a communications satellite to serve USA, Japan and Australia at the same time?

2. Are there markings on your globe that show the ground traces for equatorial and polar orbits? If so, identify them in your report. If not, use a roll of removable tape to mark them on the globe.

3. Use a roll of removable tape to mark the ground trace of an orbit inclined by 45 degrees. Start by using a protractor or 45° triangle and place the tape on the globe so it crosses the equator at 45°. Stick the tape on the globe to complete an orbit, making sure the tape is straight and not stretched. This ground trace should reach a maximum of latitude of 45° north and $45°$ south, and it should close on itself. Ground traces for satellites in such orbits do not actually close on themselves because of Earth's rotation, as discussed above.

C. Questions and Problems

1. Determine the number of sunrises and sunsets that astronauts would observe during a 24 hour day from a crewed satellite on LEO.

2. Calculate the velocity and period of a low lunar orbit, given that the Moon's mass and radius are $7.35 \cdot 10^{22}$ kg and 1740 km, respectively.

3. Given the idea that a smaller ΔV, and therefore less fuel, is required for launching from the equator of Earth, why did the Apollo astronauts land close to the lunar equator? Consider the ΔV required for both the landings and subsequent take-offs.

The Elliptical Orbit of Mercury

Planetary orbits were discovered to be elliptical by Johannes Kepler in 1609 using the most accurate data then available on the angular positions of the planets. This data was recorded at Uraniborg, Tycho Brahe's observatory on the island of Hveen, Denmark, during the last few decades before telescopes were used in astronomy. Consequently, the accuracy of the data was limited by the approximate 1 minute-of-arc resolution of the human eye. Kepler acquired Tycho's data after his death in 1601. This exercise will do graphically what Kepler did with mathematical calculations. How can the changing positions of a planet in the night sky be used to show the planet's orbit is an ellipse? Pondering this question should help one appreciate Kepler's contribution to science. This exercise also provides an endorsement of accurate data, for with less precise data Kepler may not have been able to distinguishing between circular and elliptical orbits.

The basic properties of an ellipse are shown in Figure 19-1, where the ratio of **c** to **a** gives the **eccentricity, e**. Note that when **c** equals zero, the two foci merge into a single point, the eccentricity goes to zero, and the figure becomes a circle.

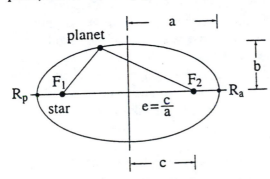

Figure 19-1: An ellipse. a is the semi-major axis, b is the semi-minor axis, c is the distance from the center to a focus, e is the eccentricity, R_p is the peri-distance, and R_a is the apo-distance.

Mercury is moderately difficult to observe, because it is relatively small, fairly dark, always close to the Sun, and, except during some total solar eclipses, is never in a dark sky. Normally, Mercury is seen either in the eastern sky in the morning twilight just before sunrise, or in the evening twilight of the western sky just after sunset. The best time to observe Mercury on each passage is when it is near its greatest angular distance from the Sun. This angle is known as the **maximum elongation**.

Tycho's records contained right ascension (RA) and declination (DEC) values for the Sun, Mercury, and many other objects. Kepler calculated the angular distance between the Sun and Mercury, which gave the values of maximum elongation for each passage of Mercury. A list of dates and values of maximum elongations are given in Table 1. Plotting these data using the procedure described below will show that the orbit of Mercury is elliptical. The eccentricity of the orbits of all other planets, excluding Pluto, are much less than that of Mercury. The graphical procedures used here would make the orbits of all other planets except Pluto appear as circles. When Kepler studied the orbits of the planets, he did not depend on diagrams, but rather used more accurate mathematical calculations which showed that the orbits of the other planets were also ellipses. See Exercise 20, Motions of Mars. His analysis of planetary orbits led to a general statement that is now known as Kepler's first law: **The orbits of planets are ellipses with the Sun at one focus**. Also see Exercise 16, Kepler's Laws of Planetary Motion I.

The ancient astronomers of Greece were familiar with the planet Mercury, but they also believed they knew Apollo, another planet. Apollo appeared only during the morning in the eastern sky. It was actually Mercury, but they did not recognize it as the same object that appeared in the evening sky. It is easy to understand how the ancients could have made this mistake; after all, east and west are opposites.

The Moon's orbit is also an ellipse. It can be easily verified that the Moon's orbit about Earth is not a circle: just photograph the Moon every few days or so over a month using the same camera and lens, and compare the image sizes. Many introductory astronomy texts include such a series of pictures to show the Moon's phases. The disk of the Moon appears its smallest each month at apogee (its farthest point from Earth) and its largest at perigee.

There are special numerical relationships between the periods of revolution and rotation of both Mercury and the Moon. The ratio for the Moon is one-to-one, as it rotates once during each revolution around Earth. Consequently, the same side of the Moon is always seen from Earth. If this relation existed between a planet and the Sun, then on that hypothetical planet there would be a year, the period of its revolution, but there would be no day, or solar day, as the Sun would never rise or set. The Sun would remain fixed at the same place in the sky all year, just as Earth remains fixed in the lunar sky when observed from the surface of the Moon.

Features on Mercury's surface are hard to observe from Earth, because they are not very large or distinctive and because Mercury is close to the Sun. The angular size of Mercury at it closest approach to Earth is only about 13 seconds-of-arc, so even with high magnification, only a few surface features are faintly discernible. For comparison, the angular size of the Moon is about half a degree, or 1800 seconds-of-arc. In the mid-nineteenth century, Giovani Schiaperelli concluded from observations that Mercury kept one side facing the Sun. It was reasonable that this would be the case, as the Sun causes large tidal forces on Mercury that could have locked its rotation and revolution into a one-to-one resonance, just like the Moon. However, in 1965, astronomers reflected radio waves off Mercury from the Arecibo radio telescope in Puerto Rico, and noticed that some of the waves were Doppler shifted towards longer wavelengths while others were Doppler shifted towards shorter wavelengths. The details of these observations were consistent with Mercury having a 58.65 day rotational period, which is not equal to its 87.97 day period of revolution. Observations of Mercury from Mariner space craft verified the 58.65 day rotation in 1974. This was clearly a triumph for radio astronomy.

Example 1: What is the ratio of Mercury's year to its sidereal day?

       The ratio of Mercury's year to its sidereal day is 87.97 / 58.65 or 1.500. Thus, they are locked in at three-to-two resonance. An exercise below shows that this means a solar day on Mercury is actually longer than a year on Mercury.

Procedures

Apparatus

millimeter ruler, and protractor.

A. Plotting Mercury's Orbit

1. Figure 19-2 gives Earth's orbit with the dates marked. Because of the small eccentricity of Earth's orbit, it appears as a circle with the Sun at one focus. Plot each of the maximum elongation values given in Table 1 as follows:

a) Draw a light and straight pencil line from the position of Earth to the Sun for each date.

b) Center a protractor at the position of Earth on each date given in Table 1, then measure, mark, and draw a heavier line so the angle from the Earth-Sun line to this new line is equal to the maximum elongation angle for that date. This new line represents the line of sight from Earth to Mercury for that date. Mercury must have been somewhere along the Earth-to-Mercury lines on the given dates, so make sure these lines extend well past the Sun. Keep your pencil sharp as you draw.

2. After lines are drawn for all of the data from Table 1, sketch a smooth curve around the Sun that just touches the heavier lines. Sketch this curve lightly at first, and then as it takes shape, make it darker, erasing and redrawing as needed. This smooth curve should contact each of the Earth-to-Mercury lines, but not cross any of them. This curve is your plot of the orbit of Mercury.

3. Determine the shape of Mercury's orbit by drawing a straight line through the largest "diameter" of Mercury's orbit that also passes through the Sun. This should be the major axis of an ellipse. Bisect this major axis to find the semi-major axis and the center of the ellipse. Draw the minor axis through the center, perpendicular to the major axis. Measure the length of each in mm and record them in a table. Also, measure the radius of Earth's orbit in mm, which is 1 AU on this diagram, and include that in your table.

4. Calculate the scale factor which will convert from mm to AU for your diagram. Convert the semi-major axis, **a**, of Mercury's orbit to AU. Does the result agree with 0.39 AU? Calculate the percentage error.

5. Use your measured value of **a** to calculate the sidereal period, **P**, of Mercury in years, using Kepler's third law (see Exercise 17, Kepler's Laws of Planetary Motion II). To convert to days, multiply the sidereal period in years by 365.25 days per year. Compare this value to the value of Mercury's 87.97 day, and calculate the percentage error.

6. The Sun lies at one of the foci of the ellipse which is Mercury's orbit. Measure the distance from the Sun to the center of the ellipse and call it **c**. Divide **c** by the semi-major axis, **a**, to get the eccentricity, **e**, of the ellipse. Compare your value of **e** for Mercury with the published value of 0.2056, and compute the percentage error.

7. Determine from your diagram the maximum and minimum possible distances between Earth and Mercury in mm. Calculate the distances these values represent in both AU and km, where 1 AU is $1.50 \cdot 10^8$ km.

8. Given that Mercury's diameter is 4880 km, calculate the maximum and minimum size Mercury will appear from Earth in radians and degrees. (Hint: the values in radians will be Mercury's diameter divided by its distance in the same units. Radians are converted to degrees by multiplying by 57.3 degrees per radian.)

B. The Relative Length of a Solar Day and Year on Mercury

Note: for this exercise, a new unit of time is defined, a Mercury minute, abbreviated Mm. One Mm is defined as one ninetieth of a Mercury year, so there are 90 Mms in a Mercury year. Notice that there are 60 Mms per sidereal Mercury day (because 58.65 Earth days times (90 Mms / 87.97 Earth days) = 60 Mms). Also, note that one quarter of a sidereal Mercury day is 15 Mms.

1. A greatly enlarged Mercury with a tower pointing toward the Sun is shown in Figure 19-3. At the base of the tower, it is solar noon on Mercury in the position shown. Sketch Mercury again each quarter of a sidereal day, which should show the tower pointing down at 15 Mms, right at 30 Mms, up at 45 Mms, left again at 60 Mms (the end of the first sidereal day), etc. Continue adding more sketches of Mercury and its tower at 15 Mms intervals until the Sun shines down the tower again. This will occur when it is again solar noon at the tower base, which will be the end of the first solar day on Mercury.

2. Explain how the length of a solar day on Mercury compares to a year on Mercury. Find and record the length of a solar day on Mercury in Earth days.

C. Additional Questions

1. Explain how the maximum elongation values would differ from those given in Table 1 if Mercury's orbit were a circle.

2. If each value of maximum elongation were uncertain by 4 degrees, what could be concluded from this data about the shape Mercury's orbit? Explain.

3. In what phase will Mercury appear at the times of maximum elongation given in Table 1?

4. In what phase will Mercury appear at its closest to Earth? In what phase will Mercury appear at its furthest from Earth?

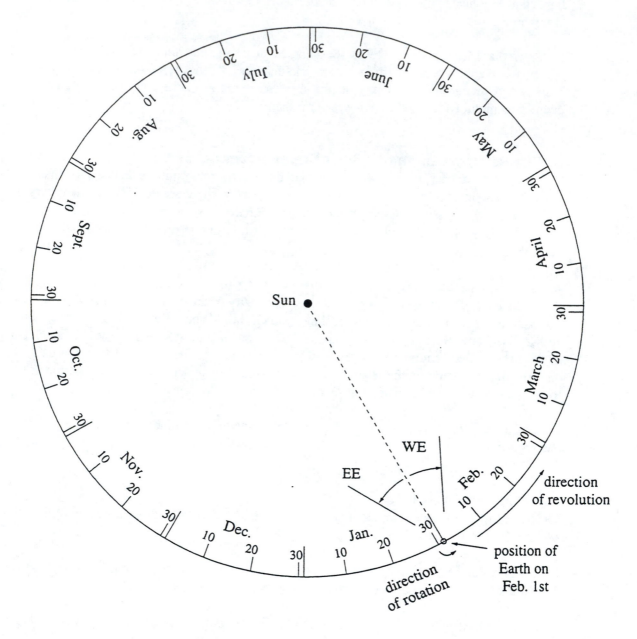

Figure 19-2: The orbit of Earth with the Sun near the center. The marks give the position of Earth in its orbit for various dates. The size of the spot representing the Sun is approximately to scale, however the line representing Earth's orbit is so thick that it would cover both the Earth and Moon in their orbits. An oversized Earth is shown for February 1, so the direction of EE (eastern elongation) and WE (western elongation) can be shown in comparison with the directions of Earth's revolution and rotation.

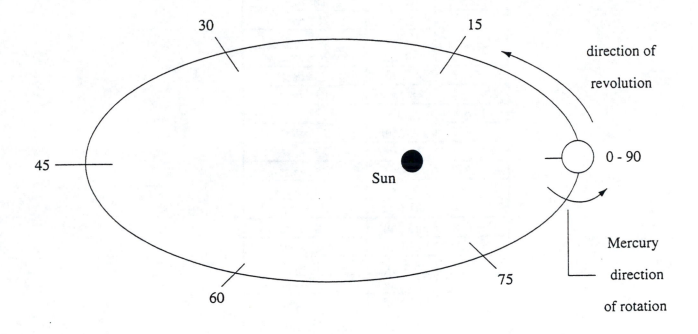

Figure 19-3: A sketch of Mercury's orbit, not to scale, showing a greatly exaggerated Mercury with a tower pointing toward the Sun. This orbit is divided into sixths, and will be used in the procedures to determine the relative lengths of a solar day and year on Mercury.

Table 1: Maximum Elongations for Mercury.

Year	Date	Angle and Direction
1967	16 Feb	18° east
	31 Mar	28° west
	12 Jun	25° east
	30 Jul	20° west
	09 Oct	25° east
	18 Nov	19° west
1968	31 Jan	18° east
	13 Mar	27° west
	24 May	23° east
	11 Jul	21° west
	20 Sep	26° east
	31 Oct	18° west
1969	13 Jan	18° east
	23 Feb	26° west
	06 May	21° east
	23 Jun	23° west
	03 Sep	27° east
	15 Oct	18° west
	28 Dec	19° east

The Motions of Mars

The daily rising, crossing the sky, and setting of the Sun, Moon, planets, and stars has been observed for all of human history. The brightness of some of these objects vary, and some exhibit curious motions relative to patterns created by others. Explaining these observations has challenged shamans, priests, and teachers for thousands of years. In the 1590's, Johannes Kepler wrote that he believed that understanding the motions of Mars was the key to understanding astronomy. This exercise describes how Kepler came to the conclusion that Mars has an elliptical orbit about the Sun (and not a circular orbit about the Earth or Sun).

Kepler found encouragement from the Copernican heliocentric model. Copernicus had learned of a Sun-centered model from the ancient Greeks and published his ideas in *De Revolutionibus*, in the year 1543 CE. Kepler was the first professional astronomer to openly uphold Copernicus' heliocentric theory. Although Kepler's work provided only a modest improvement in astronomers' ability to predict the positions of planets, it gave a rational explanation for the curious motions of planets. The exercises below show how the apparent motion of Mars results from the combination of its own and Earth's orbital motions. Kepler's accomplishments set the stage for the important work of Galileo and Newton. Also see Exercises 16 and 17, Kepler's Laws of Planetary Motion I and II.

We live in a rational age, even if many people today still do not seem rational. Many current traditions are derived from previous times when superstitions were much more influential. To illustrate, there are seven days in a week now because there were seven celestial bodies, the Sun, Moon, Mars, Mercury, Jupiter, Venus, and Saturn, seen in ancient times moving among the stars. These seven objects were all known as planets and they bear the names of mythical gods in different traditions. The names given the days of the week in European languages, listed in Table 1, illustrate connections between those names and the celestial bodies.

Table 1: Planet names for the days of the week.

Modern Day	"Planet"	Anglo-Saxon God	Day in Other Tongues
Sunday	Sun	Sunne	Sonntag (German)
Monday	Moon	Mona	Lundo (Esperanto)
Tuesday	Mars	Tiw	Martes (Spanish)
Wednesday	Mercury	Woden	Mercoledi (Italian)
Thursday	Jupiter	Thunor	Jeudi (French)
Friday	Venus	Frig	Venerdi (Italian)
Saturday	Saturn	Saturn (Roman)	Dé Sathairn (Gaelic)

The discovery of the remaining objects we know today as planets had to await the development of telescopes. Understanding that the Sun and Moon were not planets followed naturally from acceptance of the heliocentric model of the Solar System.

The planets constitute a tiny fraction of the thousands of observable sky objects. The extraordinary attention given to planets may seem reasonable when considering that many ancient peoples, lacking bright lights and spending more time outside, would have been quite familiar with the night sky. And that the human eye and brain often pay particular attention to things that move and otherwise appear unusual. The peculiar motions of the planets relative to the background stars, as can be seen on a graphical almanac and will be plotted in an exercise below, would seem to demand attention from those who noticed them.

A first-time observer would probably not notice the planets, unless the planets were brighter than other objects in the sky. And even then, he or she would not notice them moving relative to the stars. However, thoughtful observers watching the nighttime sky over many weeks would notice that the planets slowly move relative to the starry background. The usual motion of the **superior planets**, those with orbits beyond Earth's, is to move eastward in the sky relative to the stars. This is called **prograde** motion. Periodically, this prograde motion stops and a westward motion, called **retrograde** motion, occurs, followed in a few weeks by a return to prograde motion. Mars illustrates this in a clear and dramatic fashion, as indicated by its ancient Egyptian name, "sekded-ef em khetkhet," which means "one who travels backwards."

The **inferior planets**, Mercury and Venus, have their own characteristic motions and are not ordinarily seen to cross the meridian. They do cross the meridian, of course, but only when the Sun is high in the sky so they can not be seen, except when a total solar eclipse occurs near the meridian. For additional information, see Exercise 19, Elliptical Orbit of Mercury, and Exercise 24, Solar Eclipses.

Kepler worked as Tycho Brahe's assistant, starting in the year 1600. Large sextants were used in Tycho's observatory, Uraniborg, to obtain the best observational data then recorded on the angular positions of the planets. This was "pre-telescopic" data, and so was limited in accuracy by the 1 minute-of-arc resolution of unaided eyes. Kepler acquired Tycho's data upon his death, and used this data to test his ideas about the geometry of planetary orbits. Kepler knew that it took 687 days for Mars to complete one orbit about the Sun (one heliocentric circuit), while Earth requires 730 days to complete two heliocentric circuits. Kepler performed parallax measurements to determine the distance from Earth to Mars by noting the positions of Mars every 687 days, when Mars is at the same place in its orbit but Earth is not. Figure 20-1 illustrates the arrangement, and shows how it is possible to triangulate the position of Mars graphically from different positions along Earth's orbit. See Exercise 21, Parallax, for more information on Parallax. Kepler used numerical methods to determine the distance to Mars, since that could be done more accurately than graphical methods. Kepler had to interpolate to get data on some dates, as Tycho's observatory was on the island of Hveen, and data would, for example, be missing for cloudy days. By observing how Mars' distance from Earth changed, Kepler was able to show that the orbit of Mars is not circular but is close to an ellipse. Errors and uncertainties in the data limited the accuracy to which Kepler could fit Mars' orbit to an ellipse.

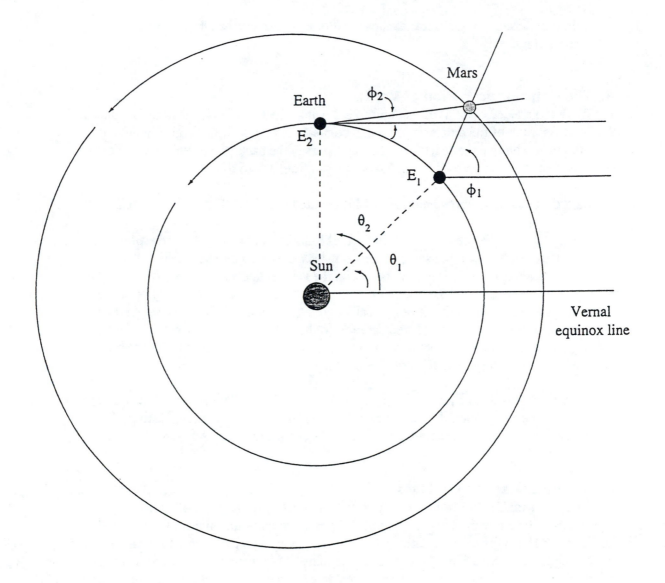

Figure 20-1: The distance to Mars from Earth at positions E_1 and E_2 is determined by triangulation. The heliocentric longitudes, θ_1 and θ_2, fix the positions of Earth along its orbit at E_1 and E_2. The geocentric longitudes of Mars, ϕ_1 and ϕ_2, determine the position of Mars when Earth is at E_1 and E_2. The orbit of Mars can then be determined by repeating this process many times. The symbols representing the Sun and planets are not to scale.

Procedures

Apparatus

drawing compass, graphical almanac and/or an observer's handbook, protractor, and straight edge.

A. The Retrograde Motion of Mars

1. Graph the trajectory of Mars in Figure 20-2 using data from Table 2. Make a point on the chart for each of the data set values and connect the points chronologically with a smooth curve. Notice that each day Mars and stars in the background will rise and set together, but the relative position of Mars changes from day to day.

2. Record the dates when the motion of Mars plotted in Figure 20-2 reverses itself.

3. Graphically explain the retrograde motion of Mars by connecting the corresponding points in Figure 20-3 with the same numbers by lines drawn to intersect the vertical line at the right. Number these intersection points on the vertical line as shown by the example. Both planets revolve about the Sun in the same direction, but Earth has a higher orbital speed and smaller orbit so that it periodically overtakes Mars. As Earth passes Mars at 5, Mars appears to move backward (retrograde), to the west in the sky, even though it is actually moving to the east in its own orbit. As Earth approaches position 7, Mars again resumes its eastward (prograde) motion in the sky.

4. Using a graphical almanac or current observer's handbook, identify the other planets that undergo retrograde motion during the current calendar year. Make a table listing these planets and the dates when the retrograde motions begin and end.

B. The Elliptical Orbit of Mars

1. Use data from Table 3 to plot the orbit of Mars relative to the orbit of Earth shown in Figure 20-4. Begin with the first data set. Measure and mark the angles of heliocentric longitude of Earth for the two dates in this data set, by using a protractor to measure the angle counterclockwise from the vernal equinox line. Mark the positions on Earth's orbit, which give the positions of Earth on the given dates. Then, construct lines parallel to the vernal equinox line which run through those positions of Earth just given. Use the protractor to measure the geocentric longitude of Mars, again in the counterclockwise direction, from the lines parallel to the vernal equinox lines. Then draw two lines from the positions of Earth in these directions (as indicated by the geocentric longitudes of Mars). These lines cross at the position Mars had on those dates, and therefore locate a point on the orbit of Mars.

2. Repeat the above for each set of data given in Table 3. Then, sketch an ellipse to provide the best fit for all of the positions determined on the orbit of Mars.

3. Draw a straight line from the apparent perihelion of Mars through the Sun to the aphelion. Bisect this line to find the center of the orbit. Measure and record the distance from the Sun to the center of the orbit (often called c). Measure and record the semi-major axis, the distance from the center of the orbit to either the perihelion position or the aphelion position. Determine the eccentricity, e, of the orbit of Mars by dividing c by the semi-major axis. Compare this value to the current value for Mars, 0.093.

Table 2: Data on the position of Mars in 1595 and 1596, at 0000 Universal Time. The date is given by month (mm) and day (dd), RA by hours (hh) and minutes (mm), and DEC by degrees (dd) and minutes of arc (mm).

Date (mmdd)	RA / DEC (hhmm (+/-)ddmm)
- 15	95 -
0620	0030 +0047
0701	0058 +0337
0710	0120 +0549
0720	0144 +0806
0801	0211 +1034
0810	0230 +1212
0820	0250 +1347
0901	0310 +1521
0910	0323 +1617
0920	0333 +1705
1001	0338 +1740
1010	0337 +1756
1020	0330 +1758
1101	0315 +1740
1110	0301 +1714
1120	0247 +1644
1201	0235 +1620
1210	0231 +1616
1220	0231 +1631
- 15	96 -
0101	0237 +1713
0110	0246 +1758
0120	0258 +1856
0201	0318 +2013
0210	0334 +2112
0220	0353 +2213

Table 3: From Tycho's data, which can be used for parallax measurements of the distance between Earth and Mars.

Date	degrees of Heliocentric Long. of Earth	degrees of Geocentric Long. of Mars
1585 Feb. 17	159	135
1587 Jan. 5	115	182
1591 Sept. 19	005	284
1583 Aug. 6	323	346
1593 Dec. 7	086	003
1595 Oct. 25	042	050
1587 Mar. 28	197	168
1589 Feb. 12	154	219
1585 Mar. 10	180	132
1587 Jan. 26	136	185

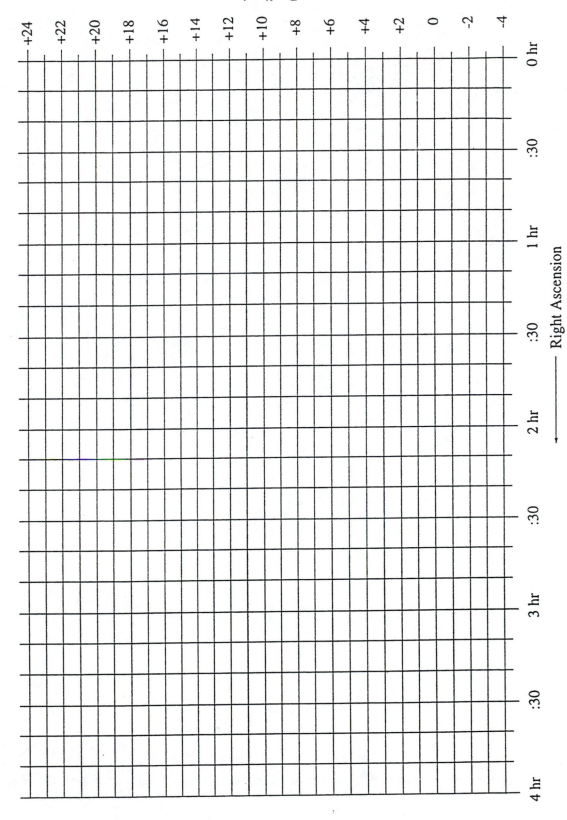

Figure 20-2: A section of sky on which to plot the motion of Mars in right ascension and declination using the data presented in Table 2.

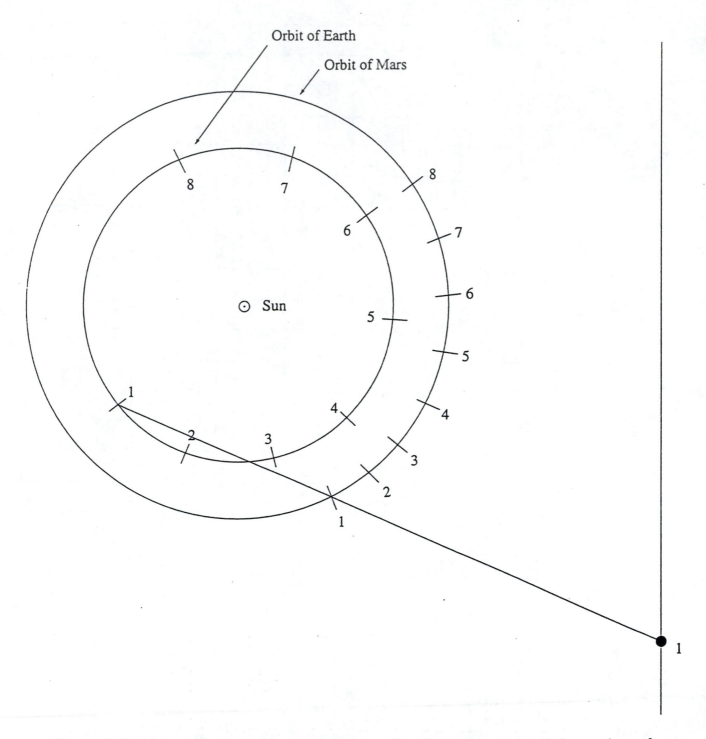

Figure 20-3: Orbits of Earth and Mars that can be used to illustrate how the relative motions of these planets result in the retrograde motion of Mars as observed from Earth.

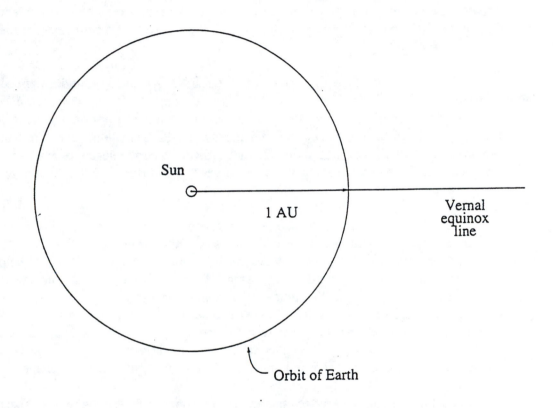

Figure 20-4: A plot of Earth's orbit about the Sun, to be used for triangulation to determine the orbit of Mars using data from Table 3.

Parallax

The distance to a star needs to be known before many of its properties can be determined. Most stars are so far away that even in the most powerful telescopes they appear only as points of light. Fortunately, there are some stars that are close enough to the Solar System that their distances can be determined by a simple geometrical method called parallax. Parallax is easy to observe: note the apparent change in position of a finger held at arm's length against the background when viewed first with one eye and then the other. This apparent movement or displacement is also detectable in stars as Earth revolves about the Sun, but only by measuring very small angles. Parallax provides a firm basis for measuring distances to nearby stars.

The word "parallax" first appears in English writings in 1594, and is defined in Webster's Ninth Collegiate Dictionary, 1983, as, "the apparent displacement or difference in apparent direction of an object as seen from two different points not on a straight line with the object." Even the parallactic shift of a nearby star due to the motion of Earth around the Sun is so small that it was not until 1838 that sufficiently large telescopes and high-quality, photographic films became available for Friedrich Wilhelm Bessel (1784 - 1846) to measure it.

This exercise demonstrates how the distance to nearby terrestrial objects can be measured using the parallax that occurs as a result of the separation of the eyes. The different view given by each eye provides important information to the brain to help produce depth perception and estimate distances, as may be noticed by using only one eye for a while. It used to be thought that humans determined the distances to objects by measuring how far the eyes had to turn inward to see the objects. This turns out to be false. If it were true, one could only determine the distance of one object at a time. A single, brief glance with both eyes open is all that is needed to estimate the distances to many objects. This is much too fast for a process that requires one to focus on each object sequentially.

Part of the ability to judge distance comes from comparing the relative sizes of familiar objects. An oak leaf, for example, looks only half as big (subtends only half the angle) when it is twice as far away. Even the blurring effect of atmospheric haze assists in distance determination and depth perception. The absence of familiar objects and an atmosphere can make it difficult to navigate on the Moon.

Parallax measurements on stars are often made by photographing the relative positions of stars about three months apart, then measuring the angle of the apparent shift. This is called heliocentric parallax with a baseline of 1 astronomical unit (1 AU), because Earth moves 1 AU perpendicular to the line of sight to the star in about three months. The baseline, b, the parallax angle, p, and the distance to the star, d, shown in Figure 21-1, are related by the parallax formula,

$$\tan(p) = \frac{b}{d}. \tag{1}$$

Notice that b and d are measured in the same units. It might be helpful if the tangent need not be determined. For small angles, p, *measured in radians*, the tangent of p is approximately equal to p. That is,

$$p \approx \tan(p) = \frac{b}{d}.$$ (2)

This is known as the small angle approximation.

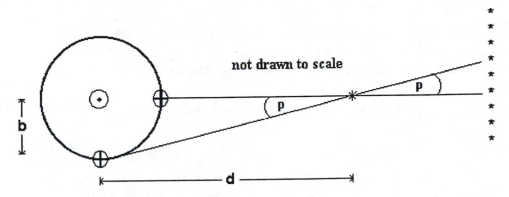

Figure 21-1: Heliocentric parallax using a baseline of 1 astronomical unit.

A new unit of distance came into usage in the early 1900's, called a parsec (pc), from "parallax of one second-of-arc." It is the distance a star must be from Earth in order to show a parallactic shift of 1 second-of-arc (1") when a baseline of 1 AU is used.

Example 1: Find the equivalent distance in astronomical units and meters for a parsec. Use Equation 1 with a parallax angle of 1 second-of-arc (1").
    Thus,

$$d = 1 \cdot pc = \frac{1 \cdot AU}{\tan(1")} = \frac{1 \cdot AU}{\tan\left(1" \cdot \dfrac{1°}{3600"}\right)} = 206{,}265 \cdot AU.$$

Since 1 AU = $1.496 \cdot 10^{11}$ m, a parsec is also

$$(206{,}265 \cdot AU)\left(\frac{1.496 \cdot 10^{11} \cdot m}{1 \cdot AU}\right) = 3.086 \cdot 10^{16} \cdot m$$

There is an inverse relationship between the parallax angle and the distance. That is, a closer object has a larger parallax angle than an object that is farther away. Since, by definition, a star that exhibits a heliocentric parallax of 1" is at a distance of 1 pc, Equation 1 can be simplified to

$$d = \frac{1 \cdot pc \cdot "}{p}.$$ (3)

Note that this equation is true only if the parallax angle is measured in seconds-of-arc and the distance is measured in parsecs.

Example 2: What will be the heliocentric parallax angle of a star that is 2 pc away.
$$p = \frac{1 \cdot pc \cdot ''}{d} = \frac{1 \cdot pc \cdot ''}{2 \cdot pc} = 0.5''.$$

Example 3: A star has a parallax angle of 0.1". How far away is it?
$$d = \frac{1 \cdot pc \cdot ''}{0.1''} = 10 \cdot pc.$$

Astronomers often use another unit, the light-year (ly). This is a unit of distance, not time, and is the distance light travels in a vacuum in one year. The exact speed of light in a vacuum is 299792458 meters per second or, to three significant figures, $3.00 * 10^8$ m/s. The number of seconds in a year is
$$1 \cdot yr = \frac{365.25 \cdot d}{yr} * \frac{24 \cdot hr}{d} * \frac{3600 \cdot sec}{hr} = 31557600 \cdot sec.$$
So, to three significant figures, a light-year is $9.46 * 10^{12}$ km. A parsec is about 3.26 light-years.

Example 4: The distance to Alpha Centauri was measured by Henderson in 1839 using parallax. Modern measurements give a parallax angle of 0.76 seconds-of-arc (4700 times smaller than one degree!). What is the distance to Alpha Centauri?
$$d = \frac{1 \cdot pc \cdot ''}{0.76''} = 1.3 \cdot pc = 4.3 \cdot ly = 270,000 \cdot AU$$
This angle, 0.76", is about the same angle subtended by an American quarter dollar coin at a distance of 8 km!

A meter cross-staff may be used to determine the distances of nearby objects using parallax. A meter cross-staff is made by taping a metric ruler (with millimeter marks) to one end of a meter stick. As in Figure 21-2, the meter cross-staff is held under the right eye so the ruler is one meter from the eye along the line of sight. An object at a distance, d, away is viewed past the ruler and is aligned with a mark on the ruler. Without moving the meter cross-staff from the right eye, vision is switched to the left eye (a baseline of y). The object appears to move against the ruler a distance of x from the original mark. Using trigonometry and similar triangles, one discovers that
$$\tan(p) = \frac{x}{d - 1 \cdot m} = \frac{y}{d} = \frac{y - x}{1 \cdot m}. \tag{4}$$
Rearranging these equations shows that d (in meters) is,
$$d = \frac{y \cdot (1 \cdot m)}{y - x}, \tag{5}$$
where the 1 meter in the numerator comes from the length of the staff.

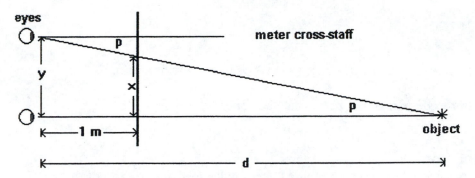

Figure 21-2: Parallax using a meter cross-staff.

The small angle approximation can be used to simplify the calculation of the parallax angle using Equation 4. If p is allowed to be measured in radians, then

$$p \approx \frac{y}{d},$$ (6)

where the distance between the observer's eyes, y, and the distance to the object, d, must be measured in the same distance units.

Example 5: Using a meter cross-staff, an object appears to shift 5.4 cm along the ruler. The distance between the observer's eyes is 6.1 cm. How far is the observer from the object? What is the parallax angle in radians?

$$d = \frac{(6.1 \cdot cm) \cdot (1 \cdot m)}{(6.1 \cdot cm - 5.4 \cdot cm)} = 8.7 \cdot m$$

$$p \approx \frac{6.1 \cdot cm}{870 \cdot cm} = 0.0070 \cdot radians.$$

Procedures

Apparatus

ruler, meter stick, and mirror (at least 15 cm long by 5 cm wide).

Note: Part A must precede part B, but part C may be done at any time.

A. Inter-eye Distance

1. Place a mirror near the edge of the lab table. Place a millimeter ruler on top of the mirror, and look at your face in the mirror. Arrange the ruler so its top edge runs from the middle of one eye to the middle of the other in your image. Close your right eye, look into the mirror and move your head (or the ruler) so you see the image of your left pupil bisected by a mark on the ruler (this may be any mark you choose). Then switch eyes without moving your head, and read the ruler mark at which you see the center of your open, right eye. The difference between these two is the distance between your eyes. It is important to measure this distance accurately and precisely (to the millimeter). Record this value.

2. Repeat the inter-eye distance measurement several times.

3. Compute the arithmetic mean of these inter-eye distance measurements. Use this mean value in the calculations of part B.

B. Parallax

Caution! Move about the lab and halls with care so as not to poke anyone in the eye with a meter cross-staff.

1. Use the meter cross-staff by holding one end close to one eye, with the ruler forming a cross-beam on the opposite end. Look at an object with one eye, noting the object's apparent position on the ruler, then switch to the other eye and again note its position. It is important that the meter cross-staff is not moved during this step. The difference between the two readings is the apparent displacement, x, due to parallax. Measure this value to the millimeter. Construct a table of the name of the object, its parallactic displacement (x), its calculated parallax angle (p in radians), its calculated distance (d_{calc} in meters), its actual distance (d_{act} in meters), and the percentage error between these two distance measurements.

2. Repeat the steps of B.1 for four more objects, as close as 2 meters and as far as 30 meters. Choose objects at distances throughout this range so you can determine for what range of distances this device is useful.

3. Discuss your results, considering the important sources of error. For what range of distances (minimum to maximum) is this method useful? Would you get more accurate results if your eyes were further apart? Why or why not?

C. Units of distance

Use a baseline of 1 AU for heliocentric parallax calculations.

1. If a star is 25 parsecs away, what is its heliocentric parallax angle in seconds-of-arc?

2. How far is the star of C.1 in light-years?

3. How long does it take light from Arcturus (Alpha Boötis) with a heliocentric parallax angle of 0.0909" to reach us? Your answer should be in years.

4. Two stars have different parallax angles. The first star has a parallax angle of 0.196", and the second star has a parallax angle of 0.0625". Which star is closer?

5. The semi-major axis of Ceres is 2.77 AU. On average, how many meters is Ceres from the Sun?

6. How long is an astronomical unit (AU) in light-minutes (lmin)? Construct a table listing the names of the planets (in increasing distance from the Sun), their mean distances in AU and in lmin.

7. Add another column to the above table that contains the semi-major axis distances of the planets in gigameters (Gm). 1 Gm is 10^9 m.

8. Approximately how long does it take a radio message to travel from Earth to Jupiter at their closest? Give your answer in minutes.

9. If Antares (Alpha Scorpii) is 120 pc away, how long would it take a spacecraft traveling at 10. percent of the speed of light to go from Earth to Antares? Give your answer in years.

10. Approximately how many centimeters is equivalent to a light-nanosecond (lns)? A nanosecond is 10^{-9} s.

Solar Lab

The Sun (or Sol) is the nearest star to Earth. In fact, it is about 300,000 times closer to Earth than the next nearest star, Proxima Centauri. It is the only star whose surface features are clearly visible from Earth, and so it provides a glimpse into the workings of other stars. The Sun's importance can not be overstressed, for without the Sun there would be no life on Earth. This lab explores some of the features and motions of the Sun.

The "surface" of the Sun can be viewed safely by projecting it onto a screen or looking through heavy filtering (such as welder's glass #14 or a special solar filter). NEVER LOOK DIRECTLY AT THE SUN, AS PERMANENT DAMAGE TO VISION MAY OCCUR! The Solar Observing Lab, Exercise 23, provides several observing exercises involving the Sun.

The Sun is not a solid object, but a huge globe of glowing plasma. The "surface" that is seen through a filtered telescope or in a solar projection is the photosphere. The photosphere is that layer of the Sun from which light seems to come. But the energy of the light originates in thermonuclear fusion reactions in the core of the Sun, where hydrogen nuclei are fused into helium. The interior of the Sun can be divided into three principal parts: the **core**, the **radiative zone**, and the **convective zone**. In the radiative zone, light is absorbed and reemitted by the solar material. The direction in which the light is reemitted is random, so it may take millions of years for light to leave the radiative zone. At the outer boundary of the radiative zone, however, the density of the solar material is low enough for convection to occur. Material that absorbs the light is heated, rises, transmits its heat to higher materials, and falls to be heated again. The convective zone consists of loops of rising and falling material. At the photosphere, the density is low enough that the hot, rising material radiates its heat away as light then sinks to be reheated.

The Sun's rotational period is not uniform across its surface; material near the equator moves about the solar axis more rapidly than material near the poles. This is called **differential rotation**. The sidereal rotational period of the Sun increases from about 25 days near the equator to about 34 days near the poles. However, since Earth revolves about the Sun at about 1 degree per day in the same direction that the Sun rotates, the synodic period is longer. As viewed from Earth, material near the Sun's equator takes about 27 days to complete one rotation, while material near the solar poles takes about 38 days.

The solar rotational period at the latitude of a sunspot may be determined by measuring the sunspot's change in solar longitude (ΔL) over a number of days (ΔT). The Sun rotates 360° in one period. Thus, the solar rotational period (P) at the latitude of the sunspot is given by

$$P = \Delta T \cdot \left(\frac{360°}{\Delta L} \right).$$
(1)

Example 1: A sunspot is seen to move 24° in 2.0 days. What is the period of rotation for material at the same latitude as the sunspot?

$$P = (2.0 \cdot d)\left(\frac{360°}{24°}\right) = 30 \cdot d$$

Figure 22-1 illustrates the Solar Angle Device (SAD). The SAD is designed to facilitate recording the Sun's movements in the sky. It consists of a flat surface (the recording platform) mounted with a perpendicular mast. The mast may be topped by a horizontal plate with a 'V' shaped cut-out. The 'V' points toward the mast. If the 'V' is placed far enough from the mast, its shadow can be seen from the equator at solar noon on both of the solstices without reorienting the SAD. The SAD is oriented so that the recording platform is level with the ground and the shadow of the 'V' falls on the surface. The recording platform is covered with paper so a new record of the Sun's motions can be made each year. A mark is placed directly below the point of the 'V' to indicate the location of the zenith as seen from the tip of the 'V'. Also, it is convenient to mount the SAD at exactly the same location and in the same orientation each time it is used.

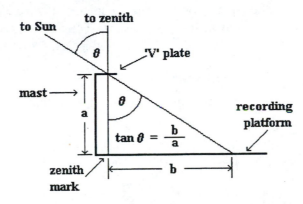

Figure 22-1: Solar Angle Device (SAD) seen edge-on.

The length of the shadow cast by a vertical pole can be used to determine the Sun's position in the sky. Solar noon is the time when the Sun is on the meridian. The Sun is highest in the sky at this time and shadows are shortest. During the day, the shadow of the 'V' is seen to move across the surface of the SAD. If this shadow is marked every few minutes (and the time of each mark recorded), the solar noon time can be determined later by finding the time when the shadow was shortest. The angle (θ) between the Sun and the zenith is calculated using

$$\theta = \arctan\left(\frac{b}{a}\right), \tag{2}$$

where a is the height of the vertical mast and b is the distance of the shadow of the 'V' from the zenith mark.

Example 2: The shadow cast by the Sun of the 'V' atop a 50. cm high vertical mast is 20. cm north of the zenith mark at solar noon. What is the altitude and azimuth of the Sun? The angle of the Sun from the zenith is

$$\theta = \arctan\left(\frac{20\cdot cm}{50\cdot cm}\right) = 22° .$$

Since this shadow is measured at solar noon and the shadow is north of the zenith mark, the azimuth of the Sun is 180° (i.e., the Sun is south of the zenith). The zenith is +90° in altitude, so the Sun must be at +68° in altitude.

The Sun's position in the sky depends on the location of the observer, the time and date, and the axial tilt of Earth. Earth's rotational axis tilts with respect to the axis of the plane of its orbit (the ecliptic) by about 23.5°. On the summer solstice, the Sun appears overhead at solar noon only for people located at 23.5°N. People north of this latitude see the Sun south of the zenith and people south of this latitude see the Sun north of the zenith. After the summer solstice, the Sun's position at solar noon appears to move southward each day. On the autumnal equinox, the Sun is overhead as viewed from the equator at solar noon. On the winter solstice at solar noon, the Sun is 23.5° south of the plane of Earth's equator and can be seen at the zenith from 23.5°S latitude. The Sun then appears to move to the north again, passing the equator on the vernal equinox. Figure 22-2 shows the changing position of the Sun in the sky during the year. The Sun is closest to the north celestial pole on the summer solstice (Figure 22-2.a) and is farthest on the winter solstice (Figure 22-2.b).

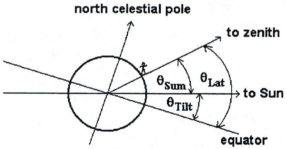

Figure 22-2(a): The position of Sun on the summer solstice.

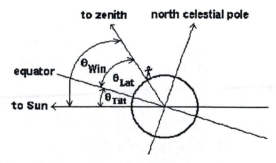

Figure 22-2(b): The position of Sun on the winter solstice.

The angular distance between the Sun and the zenith for a given location can be measured using the Solar Angle Device, but the latitude of the location and the axial tilt of Earth can not be measured directly. However, these values can be determined from the position of the Sun in the sky. At solar noon on the summer solstice (Figure 22-2.a), the Sun's angular distance from the zenith (θ_{Sum}) is given by

$$\theta_{Sum} = \theta_{Lat} - \theta_{Tilt}, \tag{3}$$

where θ_{Lat} is the latitude angle of the observer's location and θ_{Tilt} is the axial tilt of Earth. On the winter solstice (Figure 22-2.b), the Sun's angular distance from the zenith at solar noon (θ_{Win}) is given by

$$\theta_{Win} = \theta_{Lat} + \theta_{Tilt}. \tag{4}$$

Solving these two equations simultaneously, gives

$$\theta_{Lat} = \frac{\theta_{Win} + \theta_{Sum}}{2} \tag{5}$$

and

$$\theta_{Tilt} = \frac{\theta_{Win} - \theta_{Sum}}{2}. \tag{6}$$

Example 3: The angular distance of the Sun from the zenith is measured at solar noon on the winter solstice and on the summer solstice to be 58.5° and 11.5°, respectively. The Sun was south of the zenith on both occasions. Determine the latitude from which the measurements were made and the axial tilt of Earth.

$$\theta_{Lat} = \frac{58.5° + 11.5°}{2} = 35.0°$$

$$\theta_{Tilt} = \frac{58.5° - 11.5°}{2} = 23.5°.$$

Thus, Earth's axis tilts 23.5° and these measurements were taken from a latitude of 35.0°N.

The time of solar noon is different for different dates and for different locations on Earth. The daily variation in the time of solar noon is expressed by the equation of time (EOT). The equation of time gives the number of minutes that the Sun is early or late in reaching the meridian. Three factors affect the equation of time: the position of Earth along its elliptical orbit, the axial tilt of Earth, and the location of the observer.

If Earth had no axial tilt and traveled in a circular orbit, solar noon would always occur at 1200 hours for a location on a time zone line. However, Earth travels in an elliptical orbit about the Sun. Figure 22-3 illustrates the effect of an elliptical orbit on the equation of time while ignoring the effects of Earth's axial tilt and the latitude and longitude of the observer. On the winter and summer solstices, solar noon occurs at 1200 hours, just as it would if the orbit were circular. But, between the solstices, the Sun reaches the meridian earlier or later than expected. On the vernal equinox, for example, the zenith at the equator points towards the center of the orbit at 1200 hours. But the Sun is not at the center of the orbit, so Earth must rotate an additional amount in order to bring the Sun to the meridian, making the Sun late. At the autumnal equinox, Earth has already rotated past the Sun by 1200 hours, making the Sun early. If the elliptical nature of Earth's orbit were the only effect on the equation of time, it would be zero on the solstices and have only positive values between the winter solstice and the summer solstice and only negative values between the summer solstice and the winter solstice.

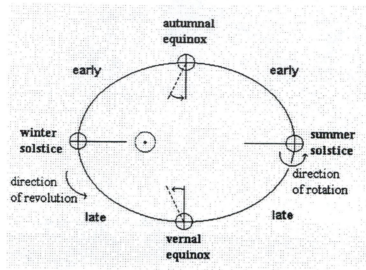

Figure 22-3: Earth's elliptical orbit affects the time of solar noon.

The axial tilt of Earth also adds to the equation of time. On the solstices, the Sun moves parallel to the equator at (360° / 365.25 d =) 0.9856° per day towards the east. But at other times of the year, the Sun does not move parallel to the equator. For example, on the vernal equinox, the Sun moves northward as well as eastward. Although the total motion of the Sun in a day is still 0.9856°, its motion eastward is slightly less. Another effect of the axial tilt also involves the motion of the Sun across the celestial sphere. Lines of right ascension are closer together further from the celestial equator. On the solstices, the Sun is furthest from the celestial equator. So, although the rate the Sun moves along the ecliptic is the same, its right ascension changes by a greater amount on the solstices than on the equinoxes. These two effects, the eccentricity of Earth's orbit and Earth's axial tilt, together are responsible for the Equation of Time. A graphical representation of this equation, called the **analemma**, is often plotted on globes of Earth and appears as an asymmetrical figure eight.

The longitude of the observer also determines the time of solar noon. If the Sun is on the meridian at one longitude, it would be east of the meridian for positions west of that longitude and west of the meridian for positions east of that longitude. Thus, solar noon occurs later for positions west of a given longitude. The number of minutes later can be computed from the difference in longitude and the rate of Earth's rotation.

Example 4: If solar noon occurs at 1228 for a particular location on the same day that solar noon occurs at 1200 hours Central Standard Time (CST), what is the longitude of this location? This is 28 minutes later than expected, so the location must be 28 minutes worth of Earth's rotation west of the time zone longitude (90°W for CST). Earth rotates 1° every 4 minutes, so

$$28 \cdot \min\left(\frac{1°}{4 \cdot \min}\right) = 7° .$$

Thus, the location has a longitude of 97°W.

Procedures

Apparatus

    drawing compass, protractor, Solar Angle Device (SAD), globe,
series of photographs of the Sun, and data from the solar observing exercise,
photocopy Figure 22-4, expanded or reduced to the size of Sun's image, on
transparency, if possible.

The following procedures (parts A, B, and C) can be done in any order.
Note: Parts A, B, and C of this lab use data collected with the Solar Angle Device during
Exercise 23, Solar Observing.

A. Latitude and Earth's Axial Tilt

1. Measure the height of the vertical mast of the Solar Angle Device to the nearest
millimeter.

2. Determine the minimum distances from the center of the base of the vertical mast to the
shadow marks for the winter and summer solstices and the vernal (or autumnal) equinox.
Compute the angular distances of the Sun from the zenith for each of these three dates.
Place all data and results in a table.

3. Determine the latitude of the SAD and the axial tilt of Earth.

4. Since the Sun is directly above the equator at local solar noon on the equinoxes, the
Sun's angle from the zenith at this time is equal to the latitude of the observer. Compare the
angle of the Sun from the zenith at local solar noon on an equinox (determined in A.2) to the
latitude computed in A.3.

B. Longitude

1. From the data recorded on the SAD, determine the time of solar noon on a given day.

2. For this same day use the almanac to determine the expected time of solar noon for the
location of the SAD.

3. Determine the longitude of the SAD from the above data.

4. Create a table of solar noon times for five dates during the year using the SAD. Use the
almanac to determine the times of solar noon for those same dates, and place these times in
the table. Determine the difference (in minutes) between the SAD and the almanac.

5. How accurate is the Solar Angle Device?

C. Sun's Rotational Period

1. Determine the solar rotational period using a series of photographs of the Sun taken over a period of about a month showing a sunspot (or sunspot group) going all the way around the Sun. Approximate the solar latitude of the sunspot. Does the measured rotational period of the Sun for the sunspot latitude agree with the expected value?

2. In Exercise 23, the Sun was sketched or photographed over a number of days. Select two sketches or photographs separated by at least two days that show a sunspot moving across the Sun's disk. Record the dates of these images.

3. Figure 22-4 contains a coordinate grid that may be used to represent latitude and longitude lines on the Sun. Carefully transfer the sunspot from the first sketch or photograph to the coordinate grid. Label the spot on the grid with a '1'.

4. Align the sketch or photograph of the Sun with the coordinate grid so the sunspot appears to move parallel to the latitude lines on the grid. Transfer the sunspot on the later sketch or photograph to the grid, and label it with a '2'.

5. Determine the number degrees of longitude between the two apparitions of the sunspot. Use the number of degrees that the sunspot moved and the number of days required for this movement to determine the period of rotation of the Sun.

6. Estimate the latitude of the sunspot used above. Is the computed period of rotation what you would expect for this latitude? Explain.

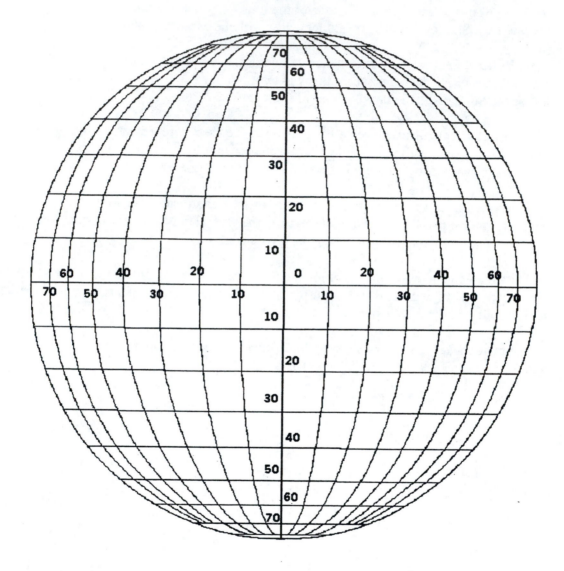

Figure 22-4: Coordinate grid showing latitude and longitude lines, for determining the rate of sunspot movements.

Solar Observing

Of the stars in the sky, only Sol (the Sun) is close enough for detailed observations of its surface. This lab explores the motion of the Sun in the sky, its rotation, its features, its light, and its composition. More exercises dealing with the Sun can be found in Exercise 22, the Solar Lab.

Sunspots and other solar features may be observed safely in two ways: by looking through a telescope equipped with a solar filter and by looking at the Sun's image projected onto a screen by a telescope or one side of a pair of binoculars. Warning: do NOT look into the projection beam, as loss of vision may result from this bright light. The Sun rotates, so tracking sunspots over a period of days can be used to determine the Sun's rotational period. Obviously, accurate sketches or photographs will be the most helpful.

The Sun's rotational period is not uniform across its surface; material near the equator moves the fastest, and the speed of rotation decreases as the solar latitude angle increases. This is called **differential rotation**. The Sun's sidereal rotational period is about 25 days near the equator, 27 days at plus and minus 30° solar latitude, and about 34 days near the poles. However, Earth moves around the Sun in the same direction that the Sun rotates. So, from Earth, the period of rotation appears longer. The period of rotation as seen from Earth is called the **synodic period**. The synodic period of rotation of material near the solar equator is about 27 days.

The synodic period of the Sun at a given latitude will be determined in this lab by measuring how far a sunspot at that latitude appears to move in a known number of days. The synodic period, S, of the Sun at the latitude of a sunspot is given by

$$S = \Delta T \cdot \left(\frac{360°}{\Delta L} \right), \tag{1}$$

where ΔT is the amount of time it takes the sunspot to move ΔL degrees across the Sun's surface.

Example 1: A sunspot appears to move 30.° across the surface of the Sun in 2.6 days. Determine the synodic period of solar rotation at the latitude of the sunspot.

$$S = (2.6 \cdot d) \cdot \left(\frac{360°}{30.°} \right) = 31 \cdot d .$$

Thus, if the sunspot lasts for 31 days, it will appear make one complete circuit of the Sun as seen from Earth.

The sidereal period of solar rotation, P, can be computed from Earth-based observations of the Sun's synodic rotational period, S, and the sidereal period of revolution of Earth, E:

$$\frac{1}{P} = \frac{1}{S} + \frac{1}{E} . \tag{2}$$

Example 2: It is observed from Earth that some sunspots require about 31 days to completely go around the Sun. Compute the sidereal period of solar rotation at the latitude of these sunspots.

$$P = \frac{E \cdot S}{E + S} = \frac{(365.25 \cdot d)(31 \cdot d)}{(365.25 \cdot d + 31 \cdot d)} = 29 \cdot d$$

Thus, at this solar latitude, the Sun takes about 29 days to complete one rotation.

During the day, the Sun moves from east to west across the sky. This motion is called **diurnal**, since it repeats every day. During the year, the Sun moves across the background of stars from west to east, passing along the ecliptic through the constellations of the zodiac. This **annual** motion is due to Earth's revolution about the Sun and repeats yearly. Since Earth's axis of rotation tilts with respect to its axis of revolution, the Sun has a north-south motion on the celestial sphere as well. This too is an annual motion of the Sun.

Solar noon is the local time when the Sun is highest in the sky. This might seem like a good time to do detailed solar observing, but instead, the best time to observe the Sun is just after dawn. During the day, sunlight warms Earth's surface. The warm ground heats the air near it. This warmer air is less dense than the cooler air around it, so it rises to higher altitudes. Since air pressure is less at higher altitudes, the air expands and cools as it rises. Cooler air is denser than warmer air, so the air sinks back to near the ground where it may be warmed again. This cycle sets up air currents that cause the image of the Sun in a telescope to flicker and wave. The image of the Sun is steadiest just after dawn, since the ground has had a chance to cool during the night. This Solar Observing lab does not require the steady atmosphere of dawn, but is designed to be done around solar noon, which is usually a more convenient time for laboratory meetings and for many of the measurements of this lab.

Accurate measurements of the Sun's position can be made using the Solar Angle Device (SAD). See Exercise 22, the Solar Lab, for more information on this device. Figure 23-1 shows the SAD as seen edge-on. The angle of the Sun from the zenith, θ, is equal to the angle formed by the mast, a line from the zenith mark to the shadow mark, and a line from the shadow mark to the 'V' plate. Since the Sun is highest in the sky at solar noon, the shadow is closest to the zenith mark at this time. Hand-angle measurements (page xi) can be used to approximate the position of the Sun in the sky.

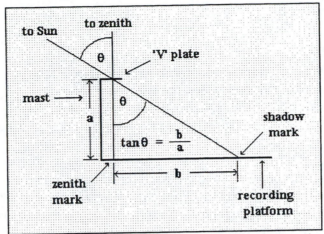

Figure 23-1: The Solar Angle Device (SAD) as seen edge-on.

It is often useful to record the Sun's position in altitude and azimuth (ALT/AZ) coordinates. Azimuth is the direction of the horizon to which the Sun is closest. This direction is represented by the angular distance clockwise along the horizon from due north. So, due north is at an azimuth of 0°, due east is at 90°, due south is at 180°, and due west is at 270°. The altitude of the Sun is the angular distance of the Sun from the closest horizon. So, if the Sun is 30° above the southern horizon, its ALT/AZ coordinates are +30°alt/180°az.

It is sometimes easier to measure the altitude of the Sun by measuring its angular distance from the zenith and subtracting this angle measure from +90°.

Example 3: At solar noon on a particular day at a particular location, the Sun is 25° due south of the zenith. What are the ALT/AZ coordinates of the Sun at this time?

The Sun is (+90° - 25°) = +65° above the southern horizon (or 180°az). Thus, the Sun is at +65°alt/180°az.

Procedures

Apparatus

telescope with projection screen, telescope with solar filter,
various eyepieces, electrical power for telescope motors,
Solar Angle Device (SAD), polarizing filters, millimeter ruler,
hand-held spectrometer, and timepiece showing correct local mean time.

This lab requires you to come to the designated solar observing area on at least two different days, some time from 1200 to 1300 hours standard time for your time zone (or 1300 to 1400 hours daylight savings time). Ideally, two (2) to five (5) days should lapse between observations. You must sign-in for each session you attend.

It may be that other students (just passing by, for example) will also want to view the Sun. You have permission to enforce a priority schedule which allows students enrolled in the lab course to dominate the telescope time, and to make markings on the SAD. It may be difficult to take enough time for careful sketches of the sunspots with people waiting to observe, but inform any guests that the telescope will be in this area on other days.

A. The Position of the Sun

1. In concert with other students, orient the SAD so that the recording platform is level with the ground and the shadow of the mast falls on the platform. Mark a double 'V' in pencil to show the area of the penumbra and umbra of the 'V' plate of the SAD every 5 minutes (or so) until it crosses the paper. Accurately record the time next to the corresponding shadow mark on the SAD.

2. Use the SAD to determine the position of the Sun in the sky at solar noon. If no shadow mark was made at solar noon but several marks are available for that day, lightly draw a line through the tips of the existing shadow marks. Since the tips of all shadow marks for this day fall on this line, interpolation can be used to fill in the times when the Sun may have been behind clouds. The shadow mark corresponding to solar noon would appear where the distance between the zenith mark and this line is shortest.

3. Determine the angular distance of the Sun from the zenith at solar noon using hand-angle measurements. It may be convenient to use a vertical line, such as the corner of a building or a light pole, as a "zenith pointer" and measure the Sun's position from that. If you are unable to make the measurement at solar noon, you must estimate where the Sun was (or will be) at solar noon and make your measurement from that position.

B. Solar Features

1. Observe the Sun using a telescope fitted with a solar filter or projection screen. If you use a solar filter, make sure it is properly placed over the objective aperture before anyone

looks through the telescope. If you use a projection screen, make sure no one looks into the projection beam.

2. On each visit to the Sun Lab, sketch and photograph (if possible) the full disk of the Sun. Show the position and relative sizes of all sunspots, faculae, and any other visible features in your sketch. Also, indicate the directions of north and south. Sketch a single sunspot cluster in detail by increasing the magnification on the telescope. Pay careful attention to the structures of the umbra and penumbra of each spot in the cluster. Record the date, time, and magnifications that you used. If you were able to photograph the Sun, also record the film speeds and exposure times that you used.

3. The approximate size of the sunspots is determined by measuring the size (in mm) of the sunspots and the Sun's image diameter (in mm) on the telescope projection screen or photograph. Make a table containing these data. Compute the actual diameter (in km) of each sunspot and place these results in the table. The Sun's actual diameter is 1,391,400 kilometers. In another column, convert these kilometer measurements into earth-diameters (1 e.d. = 12,756 km).

4. Determine the angular size of the Sun. Place the image of the Sun at the receding edge of the field of view through a telescope, and then allow it to drift out of view (if the telescope has a clock drive, the drive must be turned off). Measure how long it takes the Sun to disappear from the field of view. Repeat this measurement several times and compute the average time required for the Sun to move its own diameter (T_{avg}). Earth rotates about $1°$ every 4 minutes, so the approximate angular size, θ, is given by

$$\theta = T_{avg} \cdot \frac{1°}{4 \cdot \text{min}}. \tag{3}$$

Compute the approximate angular size of the Sun. What are some sources of error?

C. Solar Rotation

1. Estimate the number of degrees that the sunspots you sketched and photographed had moved. Also estimate the solar latitude of these sunspots.

2. Compute the synodic rotational period of the Sun at the latitude of these sunspots.

3. Compute the sidereal rotational period of the Sun at the latitude of these sunspots.

D. Solar Composition

1. Aim a hand-held spectrometer at a white object reflecting sunlight. This could be a cloud, concrete, or a piece of paper. This is a safe way to direct sunlight into the spectrometer.

2. Describe the spectrum of light that you see.

3. Record the wavelengths of the dim absorption lines, called Fraunhofer lines, in the solar spectrum. Most of these missing wavelengths of light are absorbed by substances in the Sun's chromosphere, though some are absorbed by Earth's atmosphere.

4. Use Table 1 or a similar table of the spectral lines of the elements to identify some of the substances found in the Sun's chromosphere.

Table 1: Some Fraunhofer lines.

Wavelength (nm)	Substance	Approximate Color
397	calcium	deep violet
405	mercury	violet
434	hydrogen	indigo
436	mercury	indigo
486	hydrogen	blue
517	mercury	teal
527	iron	green
546	mercury	green
577	mercury	yellow-green
579	mercury	yellow
589	sodium	yellow
590	sodium	yellow
656	hydrogen	red
687	oxygen	deep red

Solar Eclipses

A **solar eclipse** occurs when the Sun, Moon, and Earth form a straight line in space and the Moon's shadow falls on Earth. A person located on Earth may see all or part of the Sun's disk blocked from view by the Moon. A solar eclipse may be partial, annular, or total, depending on the location of the observer on Earth and through which parts of the Moon's shadow Earth moves.

By far, the most spectacular eclipse is a **total solar eclipse**, in which the Moon appears to completely cover the Sun. For such an eclipse to occur, the Moon must pass directly in front of the Sun and its apparent size must exceed the Sun's apparent size. Another way of saying this is that the **umbra** (or darkest part) of the Moon's shadow must reach Earth's surface. Only observers within the umbra, which may be up to 270 kilometers in diameter on Earth's surface, can see a total eclipse. Elsewhere, in the **penumbra** (or lighter part) of the Moon's shadow, observer's see a partial eclipse.

The motions of Earth and the Moon sweep the shadow across Earth's surface at about 1700 kilometers per hour from west to east. Using this speed and the largest umbra diameter, the longest time that a total solar eclipse can be seen from a given location is about 9.5 minutes. Typically, however, totality lasts only two to five minutes. Only observers along the narrow **path of totality** see a total solar eclipse.

If the Moon passes directly in front of the Sun but is in a part of its orbit that places it far enough from Earth that it appears smaller than the Sun, then an **annular solar eclipse** occurs. At mid-eclipse, the Sun will appear as a ring (or annulus) around the Moon.

A **partial solar eclipse** occurs when the Moon passes too far north or south of the Sun to completely cover it and also just before and after total and annular solar eclipses.

A fourth type of solar eclipse is an **annular-total solar eclipse**. Since Earth is not a perfect sphere, when the apparent sizes of the Sun and Moon are nearly the same, some observers can see an annular eclipse and some can see a total eclipse. Extremes in the Earth's terrain (such as when the path of totality crosses mountains), Earth's equatorial bulge, and changes in the Moon's distance during the eclipse can cause the Moon's apparent size to change. Some observers use aircraft to take them up into the path of totality while their colleagues on the surface view an annular eclipse.

Contact timings can assist in precisely determining the Moon's position in its orbit. Most important of these contact timings are **first contact** and **fourth contact**. They are, respectively, when the Moon first encounters the Sun's disk and when the Moon completely leaves it. **Second contact** in a total solar eclipse is when the Moon completely covers the Sun, and **third contact** is when the Moon just begins to uncover the Sun. Contact timings for features on the solar disk, such as sunspots, can be used to determine the sizes of these features.

The angular velocity of the Moon in its orbit about Earth averages about 0.5 minutes-of-arc every minute to the east. As seen from Earth, the Moon passes in front of the Sun from west to east. Since the Sun's angular diameter is about 32 minutes-of-arc, the Moon completely covers the disk of the Sun in roughly 60 minutes. The Sun's diameter is 1,391,400 kilometers, so the Moon must sweep across about 400 kilometers of the Sun's disk each second. This generalization neglects the substantial effect of the Sun's limb curvature and the eccentricities of the Moon's and Earth's orbits.

The best timing method tape records the observer's voice comments over a background of accurate timing signals, such as can be obtained from short-wave radio. The National Institute of Standards and Technology broadcasts time signals over short-wave radio station WWV from Fort Collins, Colorado and WWVH from Hawaii at frequencies of 2.5, 5, 10, 15, and 20 MHz. Similar signals may be received at frequencies of 3.330, 7.335, and 14.670 MHz over short-wave radio station CHU from Ottawa, Ontario, Canada. If such timing signals are not available, an accurate clock will suffice, but someone will have to call out the time periodically.

There are typically two varieties of electronic imaging equipment used during a solar eclipse: single image, charge-coupled device (CCD) cameras and camcorders. CCD cameras allow the user to make single or multiple exposure, high-resolution still images. These images are stored digitally either in the camera's memory or on a magnetic disk, and may be processed later by a computer. Camcorders, on the other hand, record many separate exposures sequentially on magnetic tape. These images can later be viewed as a motion picture. Many camcorders also have an audio system, which allows the user to record comments and time signals on the same tape as the images. See Exercise 13, Electronic Imaging, for more information.

Procedures

Apparatus

binoculars, telescopes, solar filters, camera, film, CCD (or camcorder), magnetic tapes, video monitors, short-wave radio capable of receiving timing signals or clock, portable tape recorder, fresh batteries, and blank tape.

Citizen Band walkie-talkies may prove useful if observing teams are set up at distant sites.

Note 1: These observing procedures are arranged below by subject. The student should read all of these procedures for general information, then highlight the appropriate procedures for the anticipated eclipse and available equipment.

Note 2: The instructor may assign students to observing teams. Teams may be responsible for different equipment or may make observations at different locations.

Note 3: The instructor may require a report detailing the eclipse. The student may wish to keep a written log of the experience. In this log may be recorded the collected data, observations, and comments.

A. Preparation (a week prior to the eclipse)

Warning: Never look directly at the Sun nor allow sunlight to enter the camera, as permanent damage may occur!

1. Discuss the upcoming eclipse, including the location of the observing area, and observing team assignments. Each team may be assigned certain duties during the eclipse or may perform observations at different sites.

2. Determine the anticipated times for your location of first contact, mid-eclipse, and fourth contact from an almanac, computer planetarium, astronomy magazine, or some other source.

3. The duration of annularity or totality is often published in almanacs and astronomy magazines and can be determined with computer planetaria. If the solar eclipse is annular or total, record the anticipated length of annularity or totality.

4. With help from the instructor, make a list of the equipment that will be available for the upcoming eclipse observing session. Take the time necessary to understand the operation of this equipment and verify that it is functioning properly.

5. Determine exposure times required to photograph the Sun through a telescope fitted with a proper solar filter. Appropriate exposure settings can be determined by setting up the equipment and photographing the Sun at various exposures. Record the frame numbers and

exposure times you use. After the film has been developed, determine which exposure times produced the best images of the Sun.

6. Explore the features of your video camera or camcorder. Using a proper solar filter, record several images of the Sun. View these images on a monitor. Determine which settings produced the best images.

B. Observing

Warning: Never look directly at the Sun nor allow sunlight to enter the camera, as permanent damage may occur!

1. Preparation at Observing Site

a. Set up the equipment at the assigned observing site at least an hour before the eclipse is supposed to begin.

b. About half an hour before first contact, make a recording of the Sun with the CCD or camcorder. View the recording and ensure that the equipment is functioning properly. Begin recording the eclipse about 5 minutes before first contact.

c. Also about half an hour before first contact, make a brief recording of your voice and timing signals. Replay the tape and ensure that the recording equipment functions properly. Then, begin recording timing signals about five minutes before first contact.

2. Disk Contact Timings

a. First contact is often the most difficult to time, since you must anticipate where and when the Moon will first appear to touch the Sun. Make an appropriate comment, such as "first mark," on the recording at the instant first contact is observed. If you miss it, describe by how much your comment was late; this is called your **personal equation**.

b. If you are observing a total solar eclipse, make an appropriate comment, such as "second mark," when the Sun is completely covered.

c. If you are observing a total solar eclipse, make an appropriate comment on the recording when third contact occurs.

d. Make an appropriate comment on the recording at the instant of fourth contact, when the Moon has completely left the Sun.

e. Any unusual astronomical observations should be noted and described on the recording.

3. Sunspot Contact Timings

a. Choose a large sunspot or sunspot group near the middle of the solar disk, if possible.

b. Record the time of first sunspot contact with an appropriate comment, such as "first spot," when the Moon's limb first begins to cover the sunspot or sunspot group. Comments for sunspot contacts should be distinguishable from those for the whole Sun.

c. Record the time of second sunspot contact with an appropriate comment, such as "second spot," when the Moon's limb first completely covers the sunspot or sunspot group.

d. If possible, repeat the above steps at third and fourth sunspot contacts.

e. Repeat these steps for other sunspots or sunspot groups.

4. Film Photography

Warning: A proper solar filter must be used at all times when photographing Sun.

a. Take a couple of exposures prior to first contact. Record the frame numbers, exposure times, and a brief description of the image.

b. Take several exposures around each of the contact times. As the light from the Sun diminishes, the exposure time may need to be increased.

c. If a total solar eclipse is being observed, the Sun may be photographed without a solar filter ONLY during totality. Take a few exposures around mid-eclipse, then replace the solar filter before third contact. Even a tiny sliver of direct sunlight is enough to permanently damage your vision. NEVER look through a camera at the Sun without a solar filter.

d. If the eclipse is annular, then a solar filter is required during the entire eclipse. Take several exposures during mid-eclipse.

e. Multiple exposures can be taken on one frame of film to show the progression of the eclipse in one image. This is best done with a single-lens reflex camera mounted on a tripod. The camera's objective lens should be covered with a proper solar filter before being aimed at the Sun. Before the eclipse, aim the camera at the location of the Sun at mid-eclipse. Exposures should be taken every five to ten minutes, starting at least 10 minutes before and continuing at least ten minutes after the eclipse.

5. CCD Imaging

Warning: A proper solar filter must be used at all times when imaging the Sun with a CCD camera.

a. Take a couple of exposures prior to first contact. Record the exposure numbers, exposure times, and a brief description of the image. If a monitor is available, verify that sharp images are being obtained.

b. Take several exposures around each of the contact times. As the light from the Sun diminishes, the exposure time may need to be increased.

c. If a total solar eclipse is being observed, the Sun may be photographed without a solar filter ONLY during totality. Take a few exposures around mid-eclipse, then replace the solar filter before third contact. Even a tiny sliver of direct sunlight is enough to permanently damage your vision. NEVER look through a camera at the Sun without the solar filter in place.

d. If the eclipse is annular, then a solar filter is required during the entire eclipse. Take several exposures during mid-eclipse.

6. Camcorder Imaging

Warning: A proper solar filter must be used at all times when imaging the Sun with a camcorder.

If you have a wide-angle lens on your camcorder, then center the camcorder on the area where the Sun will be at mid-eclipse. The Sun should be in your field of view just before, during, and just after the eclipse. If a wide-angle lens is not available, then the camcorder must track the Sun. This can be done manually or by placing the camcorder on an equatorial telescope mount with a clock drive. The audio track of the camcorder should be set up to record time signals and any observer comments.

a. Set up the camcorder and make a test recording about half an hour before first contact. If a monitor is available, verify that sharp images are being recorded.

b. Begin recording ten minutes prior to first contact and continue recording until ten minutes after last contact.

D. Summary
You may find it useful to review the recordings and photographs of the eclipse before answering the following questions.

1. Description
    a. Describe the eclipse and observing conditions.

    b. Estimate what percentage of the Sun was eclipsed from your site at mid-eclipse.

    c. Describe any changes in the brightness of the sunlight falling on the ground.

    d. Describe any irregularities along the limb of the Moon that you saw, either during the eclipse or after reviewing the recordings. Give a brief description of their appearances, times, and their locations with respect to nearby sunspots.

2. Contact Timings
    a. How long did it take the Moon to move across the Sun's disk?

    b. Compute the average speed of the Moon across the Sun's disk in kilometers per second.

    c. In a table, arrange your sunspot contact timings by sunspot or sunspot group.

    d. Determine how long it took the Moon's limb to cover each sunspot or sunspot group.

    e. Use each of these times and the average speed of the Moon across the solar disk to compute the approximate size (in kilometers) of each sunspot or sunspot group. Record all results in your table.

    f. Earth's equatorial diameter is about 12,756 km. Determine the size of each sunspot or sunspot group in your table relative to Earth's diameter.

Lunar Lab

The Moon (or Luna) has always been a subject of human fascination. Our ancestors delighted in imagining patterns in its varied surface. When the first telescopes were pointed at the Moon, we learned that the Moon is a real place with mountains and valleys just like Earth. Since the Moon rotates (about its axis) at the same rate that it revolves (about Earth), it maintains one face toward Earth. This face is called the **Near Side**, and the side not visible from Earth is the **Far Side**.

The Near Side of the Moon provides Earth-bound viewers with a variety of features. Even with the naked eye, one can tell that the Moon has two major surface types, the dark **lowlands** and the bright **highlands**. Between 4.6 and 3.9 billion years ago, the inner Solar System underwent intense meteorite bombardment. The Moon, a world with very little erosion, still shows some scars from this period. As the rate of bombardment decreased, the lunar crust began to cool and solidify. Orientale Basin (on the western limb, about 20° south of the lunar equator) is one of the last large impacts.

Between 3.9 and 3.2 billion years ago, lunar volcanism broke through the thin crust and filled the low-lying, impact basins with dark, basaltic lava. These areas are known as the **maria**, meaning "seas," because of their appearance through early telescopes. There is, of course, no liquid water on the lunar surface. The absence of an atmosphere means that water can not exist as a liquid on the Moon. If liquid water were released on the Moon, it would immediately evaporate away. Water ice might exist on the Moon in shaded, cold craters near the lunar poles, but none has been found yet. There are other "water-like" features on the Moon, such as sini (bays), pali (marshes), and, only one, oceanus (ocean).

The highlands consist of mountains and high plains. They are the light-colored areas on the Moon, and are made mostly of anorthositic rock and breccia. Breccia consists of rock fragments fused together by the intense heat of meteorite bombardment. Lunar mountain ranges are named after mountain ranges on Earth; for example, Montes Caucasus is named after the Caucasus Mountains of Earth.

The word "**crater**" comes from the Latin, meaning cup or bowl, and refers to the bowl-like holes left in a surface after an impact. Lunar craters are named after scientists and philosophers, such as Kepler and Plato. Since the maria were formed later in lunar history than the highlands, maria have fewer craters.

Rilles come in two varieties: **sinuous** and **graben**. Sinuous rilles are narrow, winding channels cut by flowing lava. Some of these channels were formed from lava tubes whose roofs have since collapsed. Graben (straight) rilles are where sections of the crust have dropped to lower levels due to faults in the crust. Normal faulting occurred as the Moon cooled, producing features like the Straight Wall, a 110 kilometer long cliff with a 600 meter drop.

Latitude and longitude coordinates have been established for the Moon. The latitude of any location on the Moon is the angular distance north or south of the lunar equator. Lunar longitude is the angle measure east or west of the lunar prime meridian. As with the Prime Meridian of Earth, the location of the Moon's prime meridian was arbitrary. By international agreement, the lunar prime meridian splits the Near Side into equal eastern and western halves. This makes the lunar prime meridian pass through Sinus Medii (Bay of the Middle).

The Moon goes through phases as it orbits Earth. When the Moon is close to the Sun in the sky, the Near Side is not illuminated by the Sun. This phase is called **new**. The new Moon is hard to find while it is lost in the Sun's glare. However, within the next few days, the Moon can be seen again as a thin (young) crescent right after sunset. When the Moon reaches a quarter of the way around its orbit, it is said to be at the **first quarter** phase. The **terminator**, which divides the night-side from the day-side, then lies along the prime meridian. This makes the Near Side appear cut in half with the eastern half illuminated by the Sun. As the Moon moves opposite the Sun in the sky, the terminator lies along the limb (as it did at new, but now the entire Near Side is in sunlight). This is the **full** phase and might be thought of a second quarter. At three quarters of the way around its orbit, the Moon is said to be at **last quarter** or **third quarter**. The terminator again lies along the prime meridian, though the western half of the Near Side is now illuminated. Finally, the Moon returns to its new phase as the cycle begins again.

The Moon is said to be a **waxing crescent** between new and first quarter, a **waxing gibbous** between first quarter and full, a **waning gibbous** between full and last quarter, and a **waning crescent** between last quarter and new. The Moon requires about 27.3 days to orbit Earth relative to the distant stars; this is the Moon's **sidereal period**. But Earth is also moving during this time, so it takes the Moon 29.5 days to go through a complete cycle of phases; this is the Moon's **synodic period**.

The plane of the Moon's orbit about Earth is inclined to the ecliptic by about 5°. That is, the Moon's orbit takes it as far as 5° north of the ecliptic and as far as 5° south of the ecliptic. Eclipses may occur only when the Moon is crossing the ecliptic. If the inclination angle were zero, the Moon would always be on the ecliptic and there would be eclipses of the Sun and Moon every month.

Procedures

Apparatus
millimeter ruler, pencils for sketching, photographs and maps of the Moon.

A. Determining Current Conditions

1. Use the almanac to determine the current phase of the Moon.

2. Determine the rise and set times of the Moon as seen from the Astronomy Lab.

3. Use one of the above times and a star wheel to determine the constellation in which the Moon currently resides.

4. For the current lunation, determine the dates of the major phases. Start with the last new Moon and proceed through the next.

B. Learning the Lunar Features

1. Use photographs and maps of the Moon, to make a sketch of the Near Side. This sketch should show the major mountain ranges, maria, and craters.

2. Label the following mountain ranges on your sketch: Montes Apenninus, Montes Caucasus, and Montes Haemus.

3. Label the following maria on your sketch: Mare Crisium, Mare Fecunditatis, Mare Frigoris, Mare Humorum, Mare Imbrium, Mare Nectaris, Mare Nubium, Mare Serenitatis, Mare Tranquillitatis, Mare Vaporum, Oceanus Procellarum, Sinus Aestuum, and Sinus Medii.

4. Label the following craters on your sketch: Archimedes, Bullialdus, Clavius, Copernicus, Eratosthenes, Gassendi, Kepler, Plato, Ptolemaeus, Theophilus, Tycho, Walter.

C. Determining Lunar Latitude and Longitude

1. Count the number of days since the last new Moon.

2. Compute the current lunar longitude of the illuminating edge of the terminator. Remember that the illuminating edge of the terminator is at 90°E lunar longitude at new, 0° at first quarter, 90°W at full, and 180° at last quarter. Be sure to record the current date and time.

3. Determine the lunar latitude and longitude of the following craters: Archimedes, Bullialdus, Copernicus, Plato, Theophilus, and Tycho.

4. Mark your sketch with the locations of the Apollo landings. The latitude and longitude coordinates of these sites are given in Table 1.

D. Determining Actual Size

1. Determine the scale of your photograph or map of the Moon by measuring its diameter (in millimeters) and dividing it into the actual diameter of the Moon (3476 kilometers).

2. Measure the diameters (in millimeters) of the following craters on your photograph or map of the Moon: Archimedes, Copernicus, Eratosthenes, Ptolemaeus, and Theophilus.

3. Compute the actual diameters (in kilometers) of the above craters.

Table 1: The locations of the Apollo landings

Apollo Mission #	date	lunar lat./long.
11	20 July 1969	0°N/23°E
12	19 November 1969	2°S/23°W
14	05 February 1971	4°S/17°W
15	30 July 1971	24°N/2°E
16	21 April 1972	11°S/16°E
17	11 December 1972	20°N/31°W

Lunar Observing

Lunar observing is an ancient and time-honored pastime celebrated in many cultures. Ancient peoples were keenly interested in the Moon's influences on Earth: eclipses, tides, weather, animal migrations, reproductive periods, etc. It was not until the 17th century that Galileo Galilei made the first recorded telescopic observations of our natural satellite. Many introductory astronomy texts show reproductions of his sketches. See IL-LUN, Lunar Lab, for more information about our nearest celestial neighbor.

Some features visible on the Moon are: maria (pronounced MAH-ree-ah), highlands, craters, rays, and rilles. A **mare** (pronounced MAH-rey) is an ancient, dark, lava plain. The **highlands** are the light-colored regions, and are typically mountainous. **Craters** are the bowl-shaped impact sites. The highlands are much older than the maria and contain many more craters. Some craters have raised central peaks and some have rays. **Rays** are light-colored streaks that radiate away from the crater. **Rilles**, long cracks in the lunar surface, are not easy to spot, but can usually be found in large impact craters and in some maria.

Lunar observing, with the exception of observing eclipses and occultations, is best done before and around the quarter phases. It might seem like the full phase would be an ideal time to observe the Moon, since the Earth-facing side is fully illuminated. However, this is a poor time, because the Sun is high in the lunar sky and shadows are very short. Shadows give contrast and definition to the lunar features, providing a "3-D" look. Also, the full Moon is often too bright to bear observing through a large telescope without special filters.

Some of the most interesting observing conditions are near or on the lunar **terminator**, the line that divides night from day. Since the Moon orbits Earth in about 30 days, the terminator advances to the west (on the Moon) about 12° per day. By observing each night, one can view the lunar features with maximum shadow details sequentially, progressing about 12° per day across the lunar globe.

The last quarter Moon does not rise until around midnight, so evening observations of the Moon must be done near the first quarter phase. Hence, there is usually only about one week per month that is convenient for scheduling an evening lunar observing session. Daytime observations of the Moon are possible; however, the sky is bright and often requires the use of daylight filters which also dim the lunar image.

The angular diameter of the Moon can be determined from the amount of time required for its image to cross a line in a telescope fixed to Earth. Earth turns 360° in 24 hours or 1° in 4 minutes. An object on the celestial equator (with a declination of 0°) appears to move westward at that rate, and so its angular size, θ, is given by

$$\theta = T \cdot \left(\frac{1°}{4 \cdot \min} \right), \tag{1}$$

where T is the time (in minutes) required for the Moon's image to cross a line in the telescope (such as a crosshair or the receding edge of the field of view).

The angular sizes of the Sun and the Moon are both about 0.5°. This corresponds to 2 minutes of right ascension. So, when the Sun and the Moon are on the celestial equator, their images will cross a line in a fixed telescope in about 2 minutes. A **cosine correction** must be applied for objects not on the celestial equator. Thus, Equation 1 becomes

$$\theta = T \cdot (\cos\phi) \cdot \left(\frac{1°}{4 \cdot \text{min}}\right), \tag{2}$$

where ϕ is the declination angle of the object.

Example 1: Determine the amount of time required for the Moon to cross a line in a fixed telescope when it has a declination of 24° and an angular size of 0.5°.

$$T = \frac{(0.5°) \cdot \left(4 \cdot \text{min}/_{1°}\right)}{\cos(24°)} = 2.19 \cdot \text{min}$$

Thus, 2 minutes and 11 seconds are required for the Moon to move its own angular diameter.

Procedures

Apparatus

    various telescopes with various eyepieces, binoculars,
color filters, and Moon maps.

A. Pre-observing

1. Prior to observing, spend some time studying a lunar map. The lunar prime meridian (at 0° lunar longitude) is a line connecting the north and south lunar poles that passes through Sinus Medii. Also, Mare Crisium is located near the eastern limb of the Near Side, just north of the equator, which makes it an easy reference for determining the orientation of the Moon's poles.

2. Locate where the terminator is at the time of your observations by using an almanac to determine the number of days since the last new Moon, when the terminator appeared on the eastern limb. Give the lunar longitude of the terminator for the observing date and time. Remember that when the terminator is on the eastern limb, it has a longitude of 90° E.

3. Prepare a preliminary sketch of the maria, craters, and highlands you expect to observe in the observing exercise. Note that some Moon maps may be inverted to match the view through astronomical telescopes.

4. Using the almanac, determine the rise and set time of the Moon. Also, determine the constellation in which the Moon resides during the observing period.

B. Observing Exercise

1. Help set up the telescopes and other equipment. Note in your report, briefly, the help you provided.

2. Measure the ALT/AZ of the Moon at the beginning of the session. Remember to record the time, date, and your latitude and longitude.

3. Make a sketch of the entire illuminated portion of the Moon in a 10 cm diameter circle. The sketch should fill the circle. Identify the maria, major craters, and highlands. "Identify" means that you find the names of those features, and label those features on your sketch. Identify lunar north (N), lunar east (E), lunar south (S), and lunar west (W) on the sketch. Note the magnifications used for this sketch.

4. Choose one crater near the terminator to sketch in detail. Your sketch should be at least 5 cm in diameter. Show as much detail as possible in your sketch. If there are smaller craters, boulders, central peaks, or rilles inside the crater, show these. Use any magnifications you wish, but record the ones you use. Be sure to label this sketch with the name of the crater.

5. Place the image of the Moon at the receding edge of the field of view of a telescope, and then allow the Moon to drift out of view. If the telescope is equipped with a clock drive, it must be shut off. Measure how long it takes the entire Moon to disappear from the field of view. Repeat this measurement several times and compute the average time required for the Moon to move its own angular diameter (T_{avg}).

6. Determine the current declination of the Moon. The declination can be determined by measuring the Moon's angular distance from the zenith when it crosses the meridian and subtracting this distance from your latitude. This value may be available from the instructor.

7. What is the ALT/AZ of the Moon at the end of the observing period? Remember to record the time.

8. Help put away the equipment before leaving. Note on your report brief details of your participation in this effort.

C. Post-Observing Exercises

1. Determine the actual size (in kilometers) of the crater you sketched. Use a Moon map or photograph to make your measurements. The diameter of the Moon is 3476 km.

2. Compute the approximate angular size of the Moon using the Moon's current declination and the average time required for the Moon to move its own diameter.

Extra Observing Activity

Using any telescope, view the Moon through red, orange, yellow, green, and blue color filters. Filters are often used by planetary astronomers to "bring out" (increase contrast with surroundings) features on a planet's surface. Do any of the available filters bring out features on the lunar surface? If so, which filters bring out which features? It may require some time for your brain to begin seeing differences other than an overall color change.

Lunar Eclipses

A **lunar eclipse** occurs when the Moon, Earth, and Sun form a straight line in space and Earth's shadow falls on the Moon. This lab considers the lunar eclipse and provides some exercises for observing this phenomenon.

A lunar eclipse can occur only at the time of full Moon, when the Moon and Sun are on opposites sides of Earth. It would seem logical to expect a lunar eclipse at every full Moon. However, the Moon's orbit is inclined 5° to the plane of Earth's orbit, and so the Moon usually misses Earth's shadow. The Moon is most likely to pass through Earth's shadow when the full phase occurs as the Moon crosses the plane of Earth's orbit. The points of intersection between the Moon's orbit and the plane of Earth's orbit are called **nodes**.

Figure 27-1 illustrates the geometry of shadows. If the Sun were far enough from Earth to approximate a point source of light, the shadow of Earth would spread out into the Solar System, and would consist only of the dark **umbra**. The real Sun, of course, is an extended object, which covers about half a degree from Earth. This means that surrounding the umbra of Earth's shadow, there is a transition zone where the shadow becomes progressively lighter further from the umbra. This lighter part of the shadow is called the **penumbra**. As shown in Figure 27-1, the angular width of the penumbra (α) is the same as the angular width of the Sun as viewed from Earth.

Lunar eclipses are classified by which part of the shadow the Moon passes through. An **total lunar eclipse** occurs when the entire Moon passes through the umbra. When the Moon is in the umbra, it would be completely hidden from the Sun if it were not for light scattered by the Earth's atmosphere. The amount of light that reaches the Moon during this part of an eclipse depends on how much dust there is in Earth's atmosphere at the time. A **partial lunar eclipse** occurs when only part of the Moon passes into the umbra. A **penumbral eclipse** occurs when the Moon only passes through the penumbra, never reaching the umbra.

Total lunar eclipses are especially fun to watch. As the Moon is engulfed in Earth's umbra, it shines only by sunlight refracted through Earth's atmosphere. Rayleigh scattering by dust in Earth's atmosphere filters out all but the longest wavelengths of the visible spectrum, causing the Moon to appear a dark red. **Totality**, when the Moon is completely engulfed by the umbra, can last up to 104 minutes.

The darkness of the Moon at mid-totality, depends on how close to the center of the umbra the Moon passes and the amount of dust in Earth's atmosphere during the eclipse. If there have been major volcanic eruptions or forest fires prior to the eclipse, the Moon may become almost invisible in Earth's umbra. French astronomer André-Louis Danjon proposed a five-point subjective scale for evaluating the Moon's brightness at mid-totality. Table 1 lists the values of the Danjon luminosity scale.

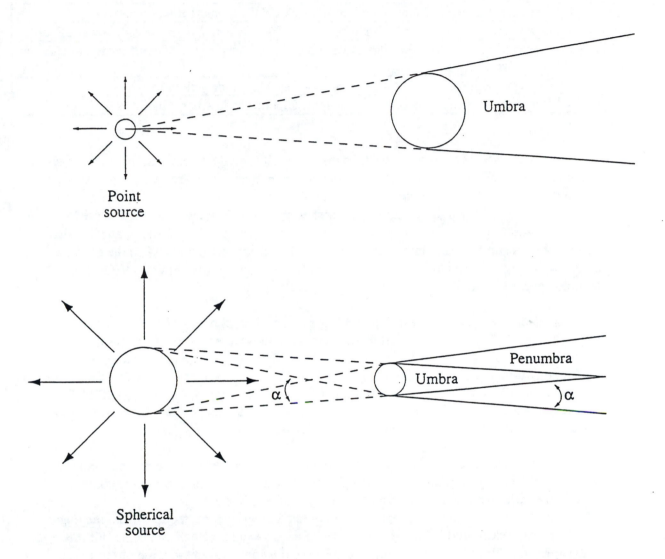

Figure 27-1: A point source, such as a distant star, casts a shadow consisting of only the umbra. An extended spherical source, such as our Sun, casts a shadow consisting of a dark inner part, the umbra, but also an annular part, the penumbra, which has the same angular width (α) as the Sun, as shown. The penumbra is not uniformly dark, but makes a smooth transition from being dark at the edge of the umbra to not being dark at all along its outside edge.

Table 1: The Danjon Luminosity Scale

Scale	Description
L = 0	Very dark eclipse. Moon almost invisible, especially at mid-totality.
L = 1	Dark eclipse, gray or brownish in coloration. Details distinguishable only with difficulty.
L = 2	Deep red or rust-colored eclipse. Very dark central shadow, while outer edge of shadow is relatively bright.
L = 3	Brick-red eclipse. Umbral shadow usually has a bright or yellow rim.
L = 4	Very bright copper-red or orange eclipse. Umbral shadow has a bluish, very bright rim.

The sequence of events of a lunar eclipse is usually noted by contacts. **First contact** occurs when the Moon begins to enter Earth's shadow. Since there are two parts of the shadow, there are two first contacts, penumbral first contact and umbral first contact. The outer penumbra is very light, so penumbral first contact is often missed by the observer. **Second contact** occurs when the Moon is completely engulfed by the shadow, and like first contact, has penumbral and umbral counterparts. **Third contact** occurs when the Moon just begins to leave the umbra or penumbra. Finally, **fourth contact** occurs when the Moon completely exits the umbra or penumbra.

Contact timings can tell astronomers a great deal about the motions of the Moon, Earth, and Sun. One of the simplest methods for making accurate contact timings uses a tape recorder and a short-wave radio receiver. The National Institute of Standards and Technology broadcasts accurate time signals over short-wave radio station WWV from Fort Collins, Colorado and WWVH from Hawaii at frequencies of 2.5, 5, 10, 15, and 20 MHz. Similar signals may be received at frequencies of 3.330, 7.335, and 14.670 MHz over short-wave radio station CHU from Ottawa, Ontario, Canada.

This method requires the observer to tape record his or her voice comments about the eclipse over a background of these time signals. For example, when first umbral contact occurs, the observer will say, "first mark," or some other pertinent comment. When reviewing the recording after the eclipse, the contact timings can be deduced from the time signals in the background. If a radio is not available, an accurate clock can be used if the time is periodically called out by another observer. If a contact is missed, the observer can make an estimate, called a **personal equation**, of how long ago the contact actually occurred.

Some observers also time contacts of the umbral shadow with specific lunar craters (see Table 2 for some prominent craters). This is made more difficult since the Sun's light strikes the Moon from overhead, producing almost no shadows to highlight the lunar features.

Table 2: Recommended Craters for Lunar Eclipse Contact Timings

1.	Grimaldi	6.	Timocharis	11.	Manilius	16.	Gassendi	21.	Billy
2.	Aristarchus	7.	Tycho	12.	Menelaus	17.	Birt	22.	Campanus
3.	Kepler	8.	Plato	13.	Plinius	18.	Abulfeda E	23.	Dionysius
4.	Copernicus	9.	Aristoteles	14.	Taruntius	19.	Nicolai A	24.	Goclenius
5.	Pytheas	10.	Eudoxus	15.	Proclus	20.	Stevinus A	25.	Langrenus

Photographing the Moon during an eclipse is a useful and rewarding way of recording observations. See Exercise 12, Astrophotography.

Procedures

Apparatus

binoculars, lunar map, portable tape recorders with tapes and fresh batteries, short-wave radio capable of receiving time signals, and telescopes with various eyepieces.

Note: Section A should be done as a class prior to the observation. Sections B, C, and D are specific to the type of eclipse that will be observed. Section E reduces the data collected in B, C, or D.

A. Pre-Eclipse (done as a class)

1. Familiarize yourself with the operation of the equipment that will be used during the lunar eclipse by operating the equipment and discussing its use with the instructor and fellow students.

2. Familiarize yourself with the lunar surface, using a lunar map. Identify craters that you plan to use for crater contact timings (see Table 2).

3. Use almanacs, computer planetaria, or astronomy magazines to find out the predicted contact timings of the eclipse. If the eclipse is total, how long is totality expected to last?

4. Considering the motions of the bodies involved, which lunar limb will encounter Earth's shadow first?

B. Penumbral Eclipse Observations

1. Help set up the equipment at the observing site. Note in your report the activities you performed. Tune in the short-wave receiver to WWV, WWVH, or CHU, make the timing signal audible, and run the tape recorders while describing the observing conditions. Play back some of the tapes to make certain all desired signals are being recorded. Start recording the timing signals at least 10 minutes prior to the predicted start of the eclipse.

2. Note when you first see the penumbral shadow. Remember that you might not see the penumbra until well into the eclipse. Which lunar limb is first contacted by the penumbra?

3. While not as spectacular as the colors of the umbra, the deeper penumbra can present a variety of dusky brown or yellowish-brown colors. Note any variations in the color of the penumbra in your report.

4. Help put away the equipment. Note in your report the activities you performed.

C. Partial Lunar Eclipse Observations

1. Help set up the equipment at the observing site. Note in your report the activities you performed. Tune in the short-wave receiver to WWV, WWVH, or CHU, make the timing signal audible, and run the tape recorders while describing the observing conditions. Play back some of the tapes to make certain all desired signals are being recorded. Start recording the timing signals at least 10 minutes prior to the predicted start of the eclipse.

2. Estimate when you first believe you see first penumbral contact. This and fourth penumbral contact are the most difficult contacts to time. Do not allow your observations to be influenced by the predicted contact times.

3. When first umbral contact occurs, call out "first mark" or some other appropriate comment to be recorded on tape along with the time signals. If necessary, state your personal equation.

4. As the umbra moves across the lunar surface, record the times of first, second, third, and fourth contacts of the craters you selected in part A.

5. The general appearance of the umbra will vary from one eclipse to the next, and may vary during the course of a single eclipse. You should note the sharpness of the umbra's leading and trailing edges frequently during the partial phases. Also, note the brightness and the color of the umbra, and note whether any variations are observed as the eclipse progresses.

6. When the umbra leaves the Moon, call out "fourth mark" or some other appropriate comment to be recorded on tape along with the time signals. If necessary, state your personal equation.

7. Help put away the equipment. Note in your report the activities you performed.

D. Total Lunar Eclipse Observations

1. Help set up the equipment at the observing site. Note in your report the activities you performed. Tune in the short-wave receiver to WWV, WWVH, or CHU, make the timing signal audible, and run the tape recorders while describing the observing conditions. Play back some of the tapes to make certain all desired signals are being recorded. Start recording the timing signals at least 10 minutes prior to the predicted start of the eclipse.

2. Estimate when you first believe you see first penumbral contact. This and fourth penumbral contact are the most difficult contacts to time. Do not allow your observations to be influenced by the predicted contact times.

3. When first umbral contact occurs, call out "first mark" or some other appropriate comment to be recorded on tape along with the time signals. If necessary, state your personal equation.

4. As the umbra moves across the lunar surface, record the times of first and second contacts of the craters you selected in part A.

5. The general appearance of the umbra will vary from one eclipse to the next, and may vary during the course of a single eclipse. You should note the sharpness of the umbra's leading and trailing edges frequently during the partial phases. Note any changes in the brightness or color of the umbra as the eclipse progresses.

6. When the Moon is completely immersed in the umbra, call out "second mark" or some other appropriate comment to be recorded on tape along with the time signals. If necessary, state your personal equation.

7. Between second and third contact, make Danjon luminosity estimates of the umbra. Make one estimate as close to mid-totality as possible, and two others shortly after second contact and just before third contact. Record the times of each estimate. These may be made on tape and the times derived later.

8. Record the times of third and fourth contacts of the craters you selected in part A.

9. When the umbra leaves the Moon, call out "fourth mark" or some other appropriate comment to be recorded on tape along with the time signals. If necessary, state your personal equation.

10. Help put away the equipment. Note in your report the activities you performed.

E. Data Reduction and Report
1. Describe the eclipse, the equipment, and observing site.

2. "Reduce" the recorded data by replaying the tapes and listing in a table the times when each important event occurred. This may require replaying the tapes several times and also using a stop watch to precisely determine the timings of the events.

3. Evaluate and interpret your experience in observing the eclipse.

4. Describe your contributions in helping set-up and pick-up equipment at the observing site.

Observing Comets

One of the most spectacular comets observed in modern times was the Great Comet of 1843. This comet was so bright that it was observed in broad daylight about one degree from the limb of the Sun. Estimates of the length of the tail ranged up to 90 degrees. Halley's comet in 1910 was visible to the unaided eye in the western sky for over a month. These apparitions have understandably created both great interest and consternation in the general public. A bright comet appeared during the Norman Conquest of England in 1066, and was soon associated with that event. In response to the Great Comet of 1456, Christian churches all across Europe added a prayer to be saved from "... the Devil ... and the comet."

The English word "comet" is derived from the Greek words "aster kometes," which mean long–haired or hairy star. Observationally, a bright comet has three main parts: a **nucleus**, a **coma**, and a **tail**. The coma surrounds the nucleus, somewhat reminiscent of a full head of unkempt hair, while the tail extends out away from the Sun. Aristotle believed comets moved within the atmosphere of Earth, while Hipparchus and others of the Pythagorean school believed comets were wanderers among the stars, much like planets. Tycho Brahe noted that a bright comet in 1577 was observed from widely separated observatories to be in the same position in the sky. Parallax would have made the comets appear at different places in the sky from different observatories, had it been closer than the Moon. This proved that this comet, and by implication, other comets, orbited beyond the Moon.

Edmond Halley's observation of the great comet of 1682 led him to research other bright comets. He noted that the great comets of 1531, 1607, and 1682 could all be one and the same if they followed a highly eccentric, elliptical orbit as allowed by Kepler's laws. In 1705, Halley predicted this comet would appear again in 1758. The comet was again seen on the evening of 25 December 1758. Although Halley had died 14 years earlier, this comet has since been known as **Halley's Comet**. The success of Halley's prediction was heralded as a great triumph for the then new scientific theories of Kepler, Galileo, and Newton. Scholars have since found references to previous sightings of Halley's Comet at least back to 240 BCE, including a description of it in 164 BCE on a Babylonian clay tablet, and a stylized image of it from 1066 on the Bayeux Tapestry. The dates of the apparitions of Halley's Comet demonstrate that the period of the comet varies from about 74 to 79 years, with the mean value being 76 years. These variations result from perturbations from the Sun, planets, and asteroids.

Example 1: Calculate the number of times Halley's Comet has orbited the Sun since it was seen in 240 BCE, assuming its period has remained 76 years. The number of years between 240 BCE and 1986 CE is 2226. Dividing this by 76 years yields, 29. This represents 30 apparitions, since it was seen in both 240 BCE and 1986.

Comets are studied by astronomers because they are one of the major types of bodies that orbit the Sun. They are intrinsically interesting, they display a wide range of physical processes, and are probably remnants of the Solar System's original material, and, hence, offer unique clues about its origin and early history. They also appear to be valuable tools for probing and monitoring the solar wind.

Newton's laws of motion and gravity predict that objects in the Solar System may move around the Sun in closed, circular and elliptical, orbits, or may pass by the Sun in open orbits, following parabolic and hyperbolic trajectories. Comets on parabolic and hyperbolic trajectories are one-time visitors to the Solar System. Figure 28-1 shows examples of these orbits. The term **periodic comet** is used to describe comets which orbit the Sun with periods less than 200 years, while comets with orbital periods predicted to be over 200 years are known as **long-period comets**. Halley's Comet is an example of a periodic comet, while the recently discovered Hale-Bopp Comet is an example of a long-period comet. It's orbital period is estimated to be over 3000 years. It is expected to reach perihelion on 31 March 1997. Orbital parameters are known for about 750 comets, of which about 600 are long-period comets. Typically, 10 new comets are discovered each year, of which only one is likely to be a bright comet.

There is a substantial body of indirect evidence that indicates comets are primarily loosely–packed conglomerates of ices, dust, and sooty, carbonaceous materials. The ices are frozen water, ammonia, methane, and carbon dioxide, all of which freeze at temperatures typical of the outer Solar System. The nuclei of comets have been described by Fred Whipple as dirty snowballs. This model helps provide a satisfying explanation of how comets appear at various points in their orbits. When a comet is in the outer Solar System, beyond the orbit of Jupiter, it is mostly a frozen nucleus covered with dust and soot, and so is small and dark and extremely hard to see. As the comet moves into the inner Solar System, its surface is warmed by the Sun. Ices near the surface sublimate to form a dilute gaseous atmosphere, the coma, that is constantly escaping the comet's weak gravity. Observations indicate the coma begins to form around 3 AU from the Sun. The gases leaving the nucleus drag loosely bound dust particles with them. As the comet moves closer to the Sun, its surface is heated more, increasing the coma's size, while the solar wind begins to blow some of the gases and dust into tails. In the inner Solar System, the coma and tail of a comet reflect about 99.9% of the light seen, though these parts only contain about one millionth of the comet's mass.

Several uncrewed spacecraft from the Soviet Union (Vega 1 and 2), Japan (Suisei and Sakigake), European Space Agency (Giotto), and USA (International Cometary Explorer) provided observations of Halley's Comet in 1985 and 1986 that are consistent with the ideas presented in this text. Some of these spacecraft even passed through the coma and imaged the 15 km long, bean-shaped nucleus. Comet IRAS-Araki-Alcock passed close enough to Earth in May 1983 to return radar signals

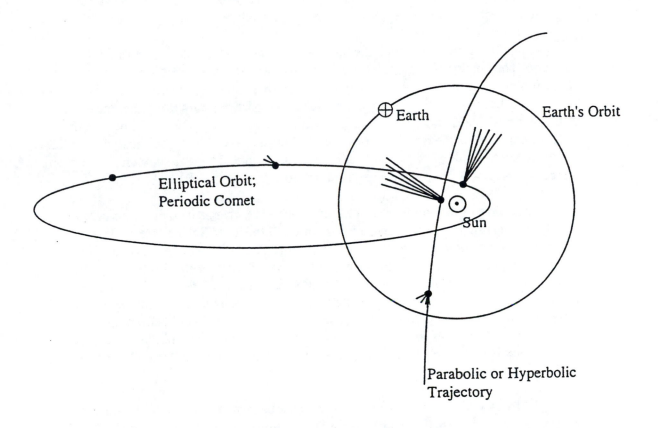

Figure 28-1: Illustrating the orbits of comets (not to scale). Note that the trails of comets point away from the Sun, due to the solar wind. Comets on parabolic or hyperbolic trajectories pass by the Sun only once.

As comets pass close to the Sun or large planets, they may be broken into pieces by gravitational forces. In 1976, the bright comet, Comet West, broke into at least four fragments as it made a particularly close pass by the Sun. In 1992, Comet Shoemaker–Levy 9 split into approximately 23 fragments when it passed close to Jupiter. This comet collided with the planet in 1994, providing a spectacular example of collisions in the Solar System. Such collisions are not all that uncommon. It is widely believed that, in 1908, a small piece of a comet exploded a few kilometers above the wilderness region of Siberia known as Tunguska. The effects of the explosion have been compared with that of a 12-megaton nuclear weapon, except without radioactive fall-out.

The dirty snowball model of comets also explains why many meteors fall in showers. The dust particles lost by comets remain in orbits with the comets around the Sun, forming trails of dust. As Earth passes through a cometary dust trail, it collides with the dust particles, producing a meteor shower as the particles burn up in the upper atmosphere. The annual August Perseid meteor shower (called "Perseid" because the meteors appear to come from the constellation Perseus) is produced by the dust trail left by comet Swift-Tuttle. There are about a dozen such annual meteor showers, some from comets whose nuclei have completely evaporated.

Quantitative development of this model indicates that as much as 1 percent of a comet's nucleus should be lost on each pass through the inner Solar System. If this is true, then comets would not be expected to last for more than a few hundred orbits. How could comets be left over from the origin of the Solar System, some 4.6 billion years ago, if they can only survive about 100 orbits? An object with a 100 year period that began orbiting the Sun 4.6 billion years ago, would have made 46 million trips through the inner Solar System by now. Yet, a comet whose period is 100 years would completely evaporate in only 10,000 years. Thus, none of the original comets should remain today. This paradox has been resolved by hypotheses advanced by Jan Oort and Gerard Kuiper and by recent observations.

In 1950, Oort hypothesized that there must be a large number of comets orbiting the Sun in a spherical shell, now called the **Oort Cloud**, at a distance of about 50,000 AU. These hypothetical comets are composed of materials left over from the original formation of the Solar System.

Example 2: Calculate the period of an Oort Cloud comet with a circular orbit of radius 50,000 AU. Recall from Kepler's third law that the square of the period in years is equal to the cube of the semi-major axis in AU. So, the period is 50,000 raised to the power of 3/2, which yields 11 million years.

Oort's hypothesis includes the idea that comets in the Oort Cloud are occasionally perturbed by passing stars, causing some to begin a long journey into the inner Solar System. Sometimes, one of these comets passes close enough to one of the planets that its orbit is perturbed into one with a shorter period, otherwise it remains a long-period comet. Since the Oort Cloud is believed to be spherical, many comets from it would have orbits out of the plane of the ecliptic. Thus, this hypothesis explains the existence of many newly discovered comets that have eccentric orbits with semi-major axes of 25,000 AU or greater and also orbit out of the plane of the ecliptic.

In 1951, Kuiper hypothesized that there is also a belt of comets orbiting from 35 to 50 AU from the Sun in the plane of the ecliptic. These comets are also remnants of the nebula that formed the Solar System. Improved electronic imaging techniques (see Exercise 13, Electronic Imaging) have brought many of these Kuiper Belt objects into the range of terrestrial telescopes. As of June 1995, 28 objects from this belt have been discovered. For example, one discovered on 28 March 1993 by Jane Luu and David Jewitt, designated 1993FW, has a diameter of 286 km, a perihelion of 42 AU, and an aphelion of 46 AU. More recent images from the Hubble Space Telescope suggest that there may be as many as 50,000 Kuiper Belt objects per square degree of the sky in the plane of the ecliptic.

There is also speculation that the Kuiper Belt may include (or is the origin of) some more familiar objects, such as the asteroid Chiron, Pluto and Charon, and, perhaps, other icy moons of other outer planets. This idea is supported by observations that Chiron suddenly increases in brightness, displaying what appears to be a coma. This could be accounted for by assuming the object has the structure of a comet and suffers occasional collisions with minor objects, releasing gas and dust.

Comet observing activities may be conveniently divided into three categories: comet hunting, faint comet observing, and bright comet observing. Comet hunting is the search for unknown comets. Successful comet hunting requires persistence, good equipment, good observing conditions, and some luck. A reward for finding a new comet is that the newly discovered comet is named after up to three of its discoverers. Comet IRAS-Araki-Alcock was the first comet named after a satellite, having been first detected by the Infrared Astronomical Satellite in 1983. The majority of comets are magnitude +12 or fainter, and observing them is challenging even with large telescopes. This exercise considers bright comet observing. Unfortunately, bright comets are relatively rare.

The tails of bright comets can often be seen to be divided into two parts, a **dust tail** and a **gas tail**. Both glow by reflected light, but the gas tail also fluoresces, converting ultraviolet light into visible light, and giving the gas tail a blue-white glow. Occasionally, small dust tails are seen that appear to point sunward, and so are called antitails. Sometimes, a comet will shed its gas tail and form another, in what is called a **gas tail disconnection event**. **Dust tail disconnection events** have also been observed. These events are probably caused by abrupt changes in the solar wind at the location of the comet.

The angular sizes of the coma and nucleus and the length of the tail can be determined by comparing them to the angular distances between nearby stars. The easiest way to make a brightness estimate of a comet is to defocus the telescope and compare the brightnesses of defocused nearby stars to the focused comet. Then, a star chart or computer planetarium can be consulted for the magnitudes of the nearby stars.

Detailed ephemerides of comet positions can be found in astronomical periodicals for most bright and periodic comets. An **ephemeris** is a list of comet positions by date. Updates on comet positions, new discoveries, observations, and images can be found at several sites on the Internet.

Photography and electronic imaging of faint comets requires techniques similar to those used for photographing deep sky objects. Photography of bright comets can be accomplished with the same techniques used in lunar and stellar astrophotography, using perhaps 2 minute exposures (see Exercise 12, Astrophotography). The eye can often perceive minute details and changes within the coma, nucleus, and tail which are missed in photographs. So, carefully prepared drawings can provide details that photographs and electronic images can not provide.

Procedures

Apparatus

comet ephemeris, orbital data on the comet, star wheel or computer planetarium, copies of a star chart (specifically, the region of the sky occupied by the comet), binoculars, telescopes, and various eyepieces.

Note: In order to observe changes in the comet's position, shape, and orientation, it is useful to divide the class into observing teams, such that each team would have responsibility for recording observations on a given night. Alternate observing sessions should be scheduled in case of inclement weather. Students should be encouraged to attend more than one observing session.

A. Pre-Observing (to be done as a class)

1. Using the ephemeris, plot the comet's positions for several days prior to, during, and after the dates of possible observing sessions on a copy of a star chart. Through which constellations will the comet pass?

2. Use a star wheel or computer planetarium to determine when the comet will be highest above the horizon. Center your observing sessions around these times.

3. Determine the comet's altitude and azimuth coordinates when it is highest in the sky during the observing sessions.

4. Use a star wheel or computer planetarium to determine the closest angular distance between the comet and Sun during the observing sessions. Comets that are too close to the Sun may be hard to see in its glare.

5. Determine the phases and positions of the Moon at the times of the planned observing sessions. What is the closest angular distance between the Moon and the comet during the observing session?

6. Use the above information and discussions with the instructor and your fellow students to plan the observing sessions and to divide the class into teams. Identify equipment that will be best suited for observing the comet during this apparition. Inspect and operate that equipment, as needed, to become familiar with its setup and use.

B. Observing (by each team)

1. Help set up the equipment at the observing site. Note in your report, briefly, the activities you performed.

2. Locate the comet through a telescope or binoculars. Record the starting time of the observations, describe the observing conditions, and the instruments and magnifications used. Record an initial description of the comet.

3. On a copy of a star chart, plot the position of the comet and the orientation of its tail.

4. Estimate the overall brightness of the comet by comparing its magnitude to that of nearby stars. Also include estimates of the magnitudes of the separate observable parts of the comet.

5. Estimate the coma's angular diameter and the angular length of the tail.

6. Make a detailed sketch of the comet. Note any unusual features.

7. Note any changes in the comet's appearance during your observation period.

8. Help put away the equipment. Comment in your report on what you did.

C. Post-Observing Exercises and Report

1. Comment on your comet observing experiences.

2. Compare the sketches you made with those made by other students and with any photographs or electronic images of the comet that are available. Note any differences in the appearance of the comet. Discuss possible reasons for these differences and record them in your report.

3. If the comet's period is known, determine during what year you may expect it to return again.

Planetary Observing

The ancients witnessed the nightly motions of the stars and perceived order in the Cosmos. But there were lights in the heavens that wandered against the background of stars. These wanderers (or **planets**) were given the names of gods, and their wanderings were believed to be portents of the future. There were seven objects known to the ancients as wanderers: the Sun, the Moon, and the five planets, Mercury, Venus, Mars, Jupiter, and Saturn. Today, we know Earth (or Terra) is also a planet. We also know of three other worlds, Uranus, Neptune, and Pluto, that were unknown to the ancients. Uranus is visible to the unaided eye, but it is so dim that the ancients never noticed it. Neptune and Pluto cannot be seen without optical aid, such as a telescope. This lab provides some planetary observing activities.

It is useful to distinguish between those planets whose orbits are inside Earth's and those whose orbits are outside of Earth's. Mercury and Venus are referred to as **inferior planets**, because their orbits keep them closer to the Sun that Earth. Mars, Jupiter, Saturn, Uranus, Neptune, and Pluto all have larger orbits than Earth and are said to be **superior planets**. Terms describing various positions of the planets in their orbits are illustrated in Figure 29-1.

When a planet appears directly opposite the Sun in the sky, it is said to be at **opposition**. Notice that only superior planets can be at opposition. This is a good time to observe a superior planet, since it is visible all night and is closest to Earth for the year. A superior planet is at **quadrature** when the angular distance between it and the Sun is 90°.

When an inferior planet is at its farthest point west of the Sun in the sky, it is said to be at its **greatest western elongation**, and it can be found low in the east just before sunrise. Likewise, when an inferior planet is at its farthest east of the Sun in the sky, it is said to be at its **greatest eastern elongation**, and it can be found low in the west just after sunset.

Inferior planets appear in conjunction with the Sun twice on each orbit, appearing in a new phase (being illuminated from behind, as is the new Moon) at **inferior conjunction** and appearing in a full phase at **superior conjunction**. The inferior planets occasionally cross the disk of the Sun at inferior conjunction, which is called a **solar transit**. Superior planets do not transit the Sun or display a new phase.

The brightness of a planet in the night sky is determined by: 1) the planet's **albedo**, which is the ratio of the amount of reflected light to the amount of incident light; 2) the planet's distance from the Sun; 3) the planet's distance from Earth, because of the $1/r^2$ reduction in the light intensity with distance; 4) the planet's size; and 5) the fraction of the illuminated part of the planet that can be seen from Earth. The inferior planets show particularly large variations in brightness, resulting from the fact that inferior planets go through a complete cycle of phases and since their distances from Earth vary by the largest fractional amounts.

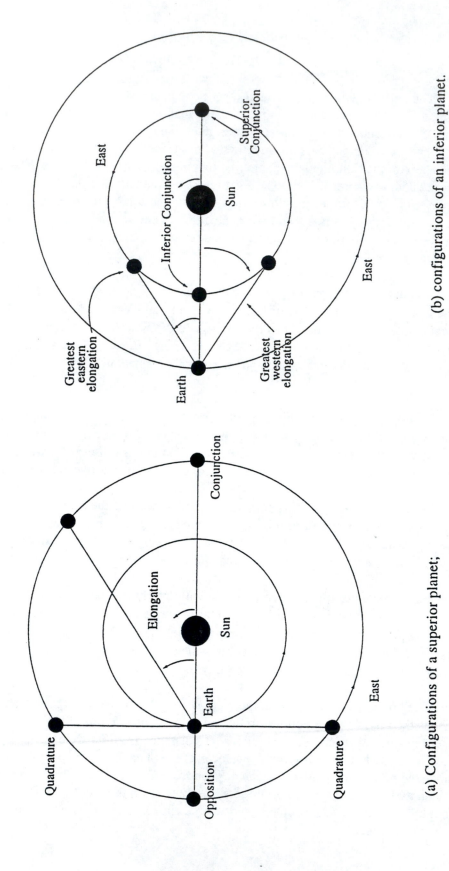

(a) Configurations of a superior planet;

(b) configurations of an inferior planet.

Figure 29-1: Various configurations of the planets in their orbits (not drawn to scale): (a) the superior planets and (b) the inferior planets.

Table 1 lists some of the characteristics of the planets. The semi-major axis of a planet's orbit is the average distance between the center of the planet and the center of the Sun, which is often measured in **astronomical units (AU)**. An astronomical unit is the average distance of Earth from the Sun or $1.496 \cdot 10^{11}$ m. The **sidereal period** of a planet's revolution is the amount of time required to complete one orbit of the Sun. The **synodic period** of revolution is the amount of time required for the planet to complete one orbit as seen from the moving Earth. The sidereal period (P) and the synodic period (S) are related by

$$\frac{1}{P} = \frac{1}{S} + \frac{1}{E}, \qquad (1)$$

where E is the sidereal period of Earth's revolution or 365.26 days.

Example 1: From Earth, it appears that Venus' period of revolution is 583.9 days. What is the true amount of time required for Venus to complete one orbit of the Sun?

Using Equation 1, the true or sidereal period, P, is

$$P = \frac{S + E}{SE} = \frac{(583.9 \cdot d + 365.26 \cdot d)}{(583.9 \cdot d * 365.26 \cdot d)} = 224.7 \cdot d$$

Table 1: Some Planetary Data, including semi-major axis, eccentricity, sidereal period of revolution, sidereal period of rotation, orbital inclination, maximum angular size as seen from Earth, equatorial diameter, mass, albedo, and number of known satellites. ('-' indicates retrograde motion.)

Name	semi-major axis (AU)	eccent.	sidereal period of revolution	sidereal period of rotation (days)	inclin. of orbit to ecliptic (degrees)	max. ang. dia. from Earth (arc-sec)
Mercury	0.39	0.206	87.97 days	58.6	7.00	12.9
Venus	0.72	0.007	224.70 days	-243.0	3.39	65.2
Earth	1.00	0.017	365.26 days	0.9973	0.00	-
Mars	1.52	0.09	686.98 days	1.026	1.85	25.7
Jupiter	5.202	0.05	11.86 years	0.41	1.31	50.1
Saturn	9.54	0.06	29.46 years	0.43	2.49	20.8
Uranus	19.19	0.05	84.01 years	-0.65	0.77	4.1
Neptune	30.06	0.01	164.80 years	0.72	1.77	2.4
Pluto	39.53	0.25	247.7 years	6.387	17.15	0.11

Name	equatorial diameter (km)	mass (kg)	albedo	num. of known satellites
Mercury	4878	$3.30 \cdot 10^{23}$	0.106	0
Venus	12,104	$4.87 \cdot 10^{24}$	0.65	0
Earth	12,756	$5.98 \cdot 10^{24}$	0.37	1
Mars	6794	$6.44 \cdot 10^{23}$	0.15	2
Jupiter	142,800	$1.90 \cdot 10^{27}$	0.52	16
Saturn	120,000	$5.69 \cdot 10^{26}$	0.47	20
Uranus	51,120	$8.68 \cdot 10^{25}$	0.50	15
Neptune	49,528	$1.02 \cdot 10^{26}$	0.5	8
Pluto	2290	$1.36 \cdot 10^{22}$	0.6	1

Kepler's First Law of Planetary Motion states that planetary orbits are elliptical with the Sun at one focus. See Exercise 16, Kepler's Laws of Planetary Motion I. Orbital **eccentricity** (e) is a measure of how far the orbit is from being a circle, and is defined as the ratio of the distance between a focus and the center of the ellipse (c) and the semi-major axis (a). Mathematically, this definition becomes

$$e = \frac{c}{a} . \qquad (2)$$

Eccentricity ranges from zero to one, where zero represents a perfectly circular orbit and one represents oscillations along a line. **Perihelion** is the point in a planet's orbit closest to the Sun, while **aphelion** is the furthest point from the Sun. Perihelion (R_p) and aphelion (R_a) distances may be calculated from the following formulae:

$$R_p = a(1 - e) \text{ and} \qquad (3)$$
$$R_a = a(1 + e), \qquad (4)$$

where a is the semi-major axis of the planet's orbit and e is its eccentricity.

Example 2: Determine the closest and furthest distances of Mars from the Sun.
   The semi-major axis of Mars' orbit is 1.524 AU and its eccentricity is 0.094.
$$R_p = a*(1 - e) = (1.524 \cdot AU)*(1 - 0.094) = 1.38 \cdot AU \quad \text{and}$$
$$R_a = a*(1 + e) = (1.524 \cdot AU)*(1 + 0.094) = 1.67 \cdot AU$$

Thus, Mars comes as close as 1.38 AU to the Sun and goes as far as 1.67 AU from the Sun.

The angular diameter of a planet is determined by both its linear (or actual) diameter and its distance from the observer. The distance to a planet may be determined in several ways, including by timing the return of radar signals bounced off the planet, by observing parallax (see Exercise 21, Parallax), and by computing the planet's orbit (see Exercise 19,

Elliptical Orbit of Mercury). Angular diameter (θ), linear diameter (D), and distance (R) are related by

$$\theta = \arctan\left(D\!\!\Big/\!\!R\right).\tag{5}$$

Example 3: On a particular day, Mars is determined to be 0.500 AU from an observer. How large does Mars appear to the observer?

0.500 AU is $7.48 \cdot 10^7$ km. And Mars' diameter is 6794 km.

$$\theta = \arctan\left(\left(6794 \cdot km\right)\!\!\Big/\!\!\left(7.48 \cdot 10^7 \cdot km\right)\right) = 0.00520° = 18.7''$$

Thus, Mars appears 18.7 seconds-of-arc across on this day.

The following paragraphs describe some salient information about the six planets that are the easiest to observe. Neptune and Pluto are difficult or impossible to observe with small telescopes, and so, are not included here.

Mercury always appears close to the Sun, mostly hidden in its glare, and is never farther from the Sun than 28°. Although Mercury is too small and far away for any features to be visible in a small telescope, it is possible to watch its phase change as it moves in its orbit. As with all planets, Mercury shines by reflecting sunlight. Mercury has an albedo of only 0.106, which means it reflects only about 11% of the incident sunlight. Its small angular size, low albedo, and close proximity to the Sun make Mercury a challenging object to observe.

Venus is second in distance from the Sun. Its greatest elongation from the Sun is 47°, which allows to be seen in the sky for at most about three hours before sunrise or three hours after sunset. Venus has a high albedo, reflecting nearly 65% of all incident light, making it at its brightest the third brightest object in Earth's sky, preceded only by the Sun and Moon. Because of its conspicuousness just after sunset or just before sunrise, it is sometimes called the "evening star" or the "morning star," depending on whether it is east or west of the Sun. Venus is said to be Earth's sister planet because of its similarity in mass and size. However, Venus' surface is totally obscured by its thick, cloudy atmosphere. Due to a run-away greenhouse effect, Venus' surface temperature reaches 750 K. Through a small telescope, Venus can be seen to go through a complete cycle of phases. This was one of the first proofs of the Copernican model of the Solar System witnessed by Galileo through his telescope.

Mars is the nearest superior planet to Earth. Mars' axial tilt is similar to Earth's, and causes it to have seasons. But Mars' seasons are twice as long as Earth's since it takes about twice as long for Mars to complete an orbit of the Sun. The changing seasons can be witnessed as the development of dust storms and the changing sizes of the polar ice caps. Mars is at its biggest and brightest, as seen from Earth, at opposition. If opposition occurs when Mars is at perihelion, Mars can come as close as 0.38 AU (or 56 million km) from Earth. Unfortunately, at this time the increased heat from the Sun often whips up planet-wide dust storms in the exceedingly thin Martian atmosphere that obscure almost all surface features.

Example 5: How much dimmer will Mars be at conjunction than at opposition, assuming the only difference is caused by the greater distance from Earth?

> Light intensity is decreased by the square of distance. At conjunction, Mars will be about 2.5 AU from Earth, while at opposition it may be as close as 0.38 AU. The variation in light intensity is therefore, $(1 / 0.38^2) / (1 / 2.5^2) = 2.5^2 / 0.38^2 = 43$. Thus, Mars at conjunction is 43 times dimmer than at opposition.

Jupiter is the largest planet in the Solar System, and it is the fourth brightest object in Earth's sky. The "surface" of Jupiter seen through telescopes is really just the cloud-tops. Dark-colored bands and light-colored zones of clouds encircle the planet. The most famous atmospheric disturbance on Jupiter is the Great Red Spot, first discovered in 1630 by the English astronomer Robert Hooke. It is a huge oval region of high pressure that rotates with a period of about seven days. Its dimensions are roughly one Earth-diameter wide by three Earth-diameters long. Jupiter's larger moons also provide interesting observing. The four Galilean moons are the largest satellites of Jupiter and were discovered by Galileo early in the 15th century. These moons appear as tiny star-like objects, forming roughly a straight line parallel to Jupiter's equator. In order of increasing distance from Jupiter, the Galilean satellites are Io (pronounce EE-oh), Europa, Ganymede, and Callisto. Ganymede has the distinction of being the largest moon in the Solar System. The daily configurations of the Galilean satellites are often published in astronomy magazines.

Saturn has been called the jewel of the Solar System. Although all of the giant planets have ring systems, Saturn's are the most extensive and the only ones that can be seen from Earth with a small telescope. The rings are made of millions of chunks of ice and rock, ranging in size from a few centimeters to tens of meters. The main ring system extends out to almost two and a half Saturn radii from the center of the planet. Saturn and its rings are tilted 26° to the plane of Saturn's orbit, so the rings change in appearance from Earth during the course of Saturn's revolution about the Sun. The rings are only a few hundred meters thick and disappear when seen edge-on. There are also cloud patterns on Saturn's "surface," but they are lighter and less distinct than on Jupiter. Titan is the largest moon of the Saturnian system and is the only moon in the Solar System with an atmosphere thicker than Earth's. Through a small telescope, it appears as a tiny, star-like object that orbits the planet approximately every 16 days.

Uranus was discovered in 1781 by the British astronomer, Sir William Herschel. By international agreement, the new planet was named Uranus, after the mythological father of Saturn. Far away from city lights, Uranus is just barely visible to the unaided eye. Even through large, Earth-based telescopes, Uranus appears only as a tiny, pale-green disk. Uranus' axial tilt is 97.92°, which means it lies essentially on its side. Its moons orbit the planet's equator, roughly perpendicular to the plane of the planet's orbit. The largest moons of Uranus, Titania and Oberon, are less than half the diameter of our Moon, and are difficult to see with small telescopes.

Procedures

Apparatus

detailed star charts or a computer planetarium program, various telescopes and eyepieces, binoculars, various color filters, and current data on the planets to be observed (including the daily configurations of the planets' moons).

Note: current planetary data can be obtained from most astronomy magazines or from a computer planetarium program.

A. Pre-Observing (to be done as a class)

All questions in this exercise pertain to the specified observing period, unless stated otherwise.

1. Use a computer planetarium program or an almanac to determine which planets will be visible and high in the sky during the observing session.

2. Obtain from the instructor, astronomy magazines or a computer planetarium the right ascension and declination (RA/DEC) coordinates, approximate distances from Earth, apparent magnitudes, and satellite configurations of the chosen planets during the observing session.

3. Use a detailed star chart (or computer planetarium) and the RA/DEC coordinates to determine the locations on the celestial sphere of the planets during the observing session. Sketch the region around each planet. Label some of the stars in your field of view with their names (if available) and apparent magnitudes. Also list the constellation in which each planet may be found.

4. Use the distances of the planets from Earth and their diameters (found in Table 1) to compute the angular diameters of the planets.

B. Observing

1. Help set up the equipment at the observing site. Note in your report, briefly, the tasks you performed. Also, briefly describe the observing conditions, noting light pollution, clouds, etc.

2. Locate each planet, using your regional sketches, if necessary.

3. Estimate the magnitude of each planet by comparing it with nearby stars.

4. Sketch each planet under low magnification, so as to include some of the neighboring stars. Show all of the planet's visible satellites in your sketch. On each sketch, note the

telescopes, magnifications, and filters you used. Describe the colors and relative brightnesses of the objects in your view.

5. Sketch each planet under higher magnification, showing as much detail as possible. On each sketch, note the telescopes, magnifications, and filters you used. Briefly describe the features that you see.

6. After the observing session, help pick up and return the equipment. Note in your report, briefly, the tasks you performed.

C. Post-Observing
1. Did your magnitude estimates (B.3) agree with the predicted values (A.2)? If not, explain.

2. If you observed a planet whose moons were visible, did your observations (B.4) agree with predicted configurations (A.2)? If not, explain. Label the moons in your sketches.

Occultations

An **occultation** occurs when one celestial body blocks another from view. Observing this can be visually exciting. The larger moons of Jupiter and Saturn are occulted almost daily by their respective planets, the Moon occults stars daily, while asteroids and other planets are involved in occultations less frequently. Several astronomy publications list predicted occultations periodically, and can be consulted to discover local opportunities for observing such events.

Many lessons have been learned from occultations. In 1675 Olaus Roemer, a Danish astronomer, noticed by timing the eclipses of Jupiter's moons, that the orbits of these moons seemed to depend on where the Earth was in its orbit. Jupiter's moons appeared early when Earth was closer to Jupiter and late when farther away. Roemer's data shows a total variation of about 16.5 minutes, which he subsequently realized was the time required for light to travel across the diameter of Earth's orbit. This provided the first accurate speed of light measurement, 2 AU per 16.5 minutes, or about 3×10^8 m/s using current values for 1 AU.

The occultation of a binary star by the Moon's limb may cut off light from one star at a time, allowing each star's brightness to be observed seperately. Thus, an occultation may also reveal that a star previously unsuspected of being a binary is in fact binary. Occultations may also provide improved information on the position and shape of the Moon, the position of a star, as well as the position, size, and shape of an asteroid. Figure 30-1 illustrates some possibilities.

A **lunar occultation** occurs whenever the Moon passes between Earth and a star or planet. The Moon rises in the east and sets in the west due to Earth's rotation. However, the Moon moves eastward relative to the background stars due to its revolution about Earth. The Moon takes about a month to complete one revolution of Earth. This is 360°/month, about 12°/day, or 0.5°/hour. Since the angular diameter of the Moon viewed from Earth is about 0.5°, the Moon moves about one lunar diameter east per hour relative to the background stars. Thus, when the Moon occults a star, the star will appear to blink off or disappear somewhere on the eastern side of the Moon, and reappear on the western side. Such an event is called a **total lunar occultation**.

Example 1: What is the maximum time a star can be blocked from view by the Moon?
Since the Moon moves its own diameter each hour, a star that is blocked from view by the widest part of the Moon will be blocked for about one hour.

North and south on the Moon are defined by the Moon's rotation. A right fist makes a handy model of the Moon. With the thumb pointing north and parallel to Earth's rotational axis, the thumb also marks the approximate position of the Moon's north polar axis. The fingers point in the direction of the Moon's motion relative to the background stars when the fist is opened, and curl in the direction of the Moon's rotation when the fist is closed.

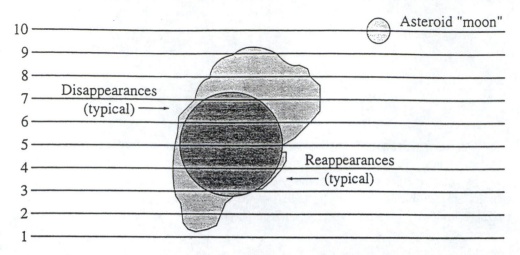

Figure 30-1: The occultation of a star by hypothetical asteroids of different shapes. The 10 lines shown represent the paths 10 different observers would see the star following during an occultation if the observers were equally spaced across the width of the occultation path. The advantage of more observers should be clear by considering what could be assumed about the shape of the asteroid from more or fewer observers. For example, consider the case where there is only one observer, say at position 5. Also, notice that timing errors distort what would be believed about the asteroids' shape. For example, if observer at position 10 has a clock that is early, the asteroid's moon may be interpreted as just a protrusion from the upper end of the asteroid.

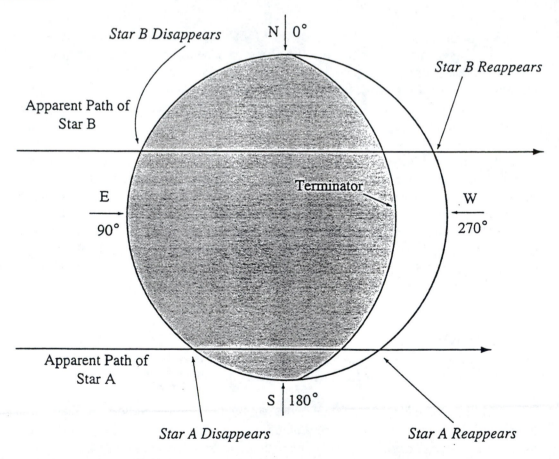

Figure 30-2: The paths of two stars showing total lunar occultations by waxing crescent Moon that is about 20% illuminated. Both stars disappear on the dark limb and reappear on the bright limb. For the disappearance of star A, the position angle is 160° and the reappearance position angle is 200°.

The positions on the Moon's limb where objects disappear and reappear during occultations are called **position angles**. Position angles are given in degrees with 0° being lunar north, 90° east, 180° south, and 270° west, as illustrated in Figure 30-2. Since it is difficult to time disappearance and reappearance of dim objects on the Moon's sunlit limb, it may be useful to know this angle prior to observing an occultation. The apparent paths of stars occulted by the Moon are shown in Figure 30-2.

A **grazing occultation** occurs when a star passes at the northern or southern poles and just appears to touch the Moon. Due to the ruggedness of the lunar surface, a star may disappear and reappear several times during the graze as it disappears behind mountains and reappears in valleys. A particular grazing occultation will only be observable from within a narrow band on Earth. If an observer is off by as little as 150 meters from the middle of this band, the star may appear to miss the Moon or be totally occulted by the Moon. Consequently, grazes are most often observed by astronomers willing to travel, and even then it is useful to have many observers spaced at intervals across the predicted band.

The term **disappearance** is used to describe when an object being occulted is no longer visible. **Reappearance** describes when the object is again visible. There may be multiple disappearances and reappearances during a grazing occultation. Disappearance and reappearance for planets require longer periods of time than for stars. **First contact** occurs when the planet's image first appears to touch the Moon. **Second contact** occurs when the planet first becomes completely hidden. **Third contact** occurs when the planet first beings to reappear. **Fourth contact** is that last apparent contact between the planet and the Moon. During a grazing occultation, multiple second and third contacts may occur as the planet's image appears and disappears in the rugged lunar surface. Figure 30-3 illustrates a presentation of hypothetical data from a grazing occultation of the Moon.

Planets are also occulted by the Moon, and these occultations may be total or partial. A **partial occultation** is a grazing occultation in which part of the occulted object remains visible. The angular size of stars are so small that observing their partial occultations are virtually impossible.

Example 2: Find the width of the band on Earth in which it is possible to observe a partial occultation of a planet and a star. Assume the planet has an angular size of 30 seconds of arc, and the star is the size of our Sun at a distance of 1 kly.

Since the distance to any planet and the star are much greater than the distance to the Moon, the angles subtended by the planet and star (α in Figure 30-4) will be very nearly the same as the width of the respective bands as seen from the Moon. See Figure 30-4. For the specified planet this angle is about 30 seconds of arc or $1.5 \cdot 10^{-4}$ radians. The width of the band for the planet is then, $w_p = 384,000$ km $* 1.5 \cdot 10^{-4}$ rad. $= 56$ km. For the star, the angle is $d_s / D_s = (14 \cdot 10^5$ km$/1000 \cdot 10^{13}$ km$) = 14 \cdot 10^{-11}$ radians, and $w_s = 384,000$ km $* 14 \cdot 10^{-11}$ rad. $= 5.4$ cm. Since the distance between the eyes of most humans and the diameter of most telescopes are greater than 5 cm, it is not realistically possible to observe partial occultations of stars 1 kly away.

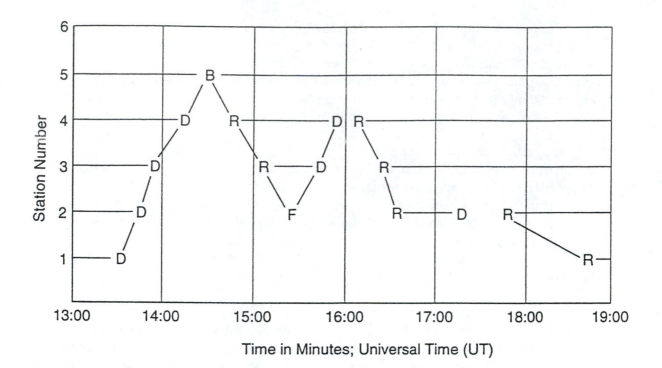

Figure 30-3; A plot of the record of six stations observing a hypothetical grazing occultation. A "D" signifies a disappearance, "R" reappearance, "B" blink, for when the star blinks out for a moment, and "F" flash, when the star flashes on for a moment.

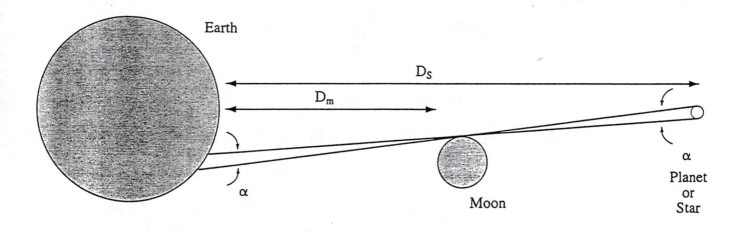

Figure 30-4: The width of the path from which a partial lunar occultation can be observed is determined by the angle α. Since the distance to the occulted object is much further from Earth than is the Moon, α is about the same as the angle subtended by the object from Earth. Distances are not to scale.

Example 3: Find the time required for disappearance when the Moon occults the planet and star considered in example 2.

The Moon moves eastward at 0.5°/hour relative to the stars, and so, after first contact, would cover the planet at about that angular rate. Thus, the planet would be covered in a time, $t_p = 30''/(0.5°/hr.) = (30''/0.5° * 3600''/°) * 3600$ seconds $= 60$ seconds. The angular size of the star is given in radians in Example 2. The star would be covered in a time, $t_s = (14 \cdot 10^{-11}$ rad. $* 57.3°/rad.)/(0.5°/hr.) = 16 \cdot 10^{-9}$ hours or 67 microseconds. Although modern electronic circuits readily accomplish accurate measurements in this time (illustrated by the fact that personal computers, operating at 50 megahertz, do 50 operations in a microsecond) the amount of light received from even bright stars is usually insufficient to allow accurate measurements of this time.

Planets and asteroids also occult stars in processes called **planetary occultations** and **asteroidal occultations**. Because of the limited angular sizes of these objects, such occultations occur with bright stars only once every few years. Occultations of dim stars occur more often but such events are not easy to observe because of the lesser amount of light received from the dim stars. Asteroids are also dim, often too faint to be seen without telescopes with apertures greater than 2 meters.

Information about planetary atmospheres can be obtained from recording how fast the star's light is dimmed by the planet. A plot of light intensity versus time is called a **light curve**. The light curve from a planetary occultation provides information about the thickness, optical density, refraction, and scattering of light by the planetary atmosphere.

Surprises may also result from observing occultations. In 1977 observers timing an occultation of the asteroid Hebe noted that Hebe occulted a star as scheduled, but they observed another disappearance and reappearance which would be consistent with Hebe having a moon. There has been no confirmation of Hebe's possible satellite, but in 1993 the spacecraft Galileo imaged asteroid 243 Ida and discovered it has a satellite, which has subsequently been named Dactyl. Perhaps, there are more satellites of asteroids waiting to be discovered.

It was predicted that Uranus would occult a star in 1977 and that this event would be visible only from a band over the Pacific Ocean. A NASA aircraft, the Kuiper Airborne Observatory, equipped with a telescope and photoelectric detectors was dispatched to the area for observing. Before the occultation was to occur, the apparatus was turned on to make sure it was recording and tracking the star correctly. This was fortunate as the star's light intensity dipped several times forty minutes before Uranus eclipsed the star, and then dipped again several times afterwards. Those dips were interpreted to be due to previously unknown rings around Uranus. The presence of the rings have since been confirmed and imaged by the Voyager II spacecraft.

The following items need to be considered to determine if observing a particular occultation is practical:

        a. the date and time of the predicted event, and the location from which it will be observable;

        b. the phase of the Moon, and magnitude and position of the star, planet, or asteroid involved; and

        c. the anticipated weather conditions.

These items provide information about where the Sun and Moon will be relative to the occultation. Daytime occultations are more difficult to observe, and are impossible for dimmer objects. Occultations close to the horizon may be obscured by clouds, dust in the atmosphere, or by objects on the horizon.

Prior to the development of modern electronics, occultations could only be observed in person and timed by hand using stop watches. Now, it is possible to record such events on video and audio tape. In the USA, the National Institute of Standards and Technology broadcasts accurate timing signals through the "WWV" short-wave radio stations at 2.5, 5, 10, 15, and 20 MHz. In Canada, timing signals are broadcast on CHU at 3.330, 7.335, and 14.670 MHz. A radio tuned to these stations will play an audio signal, which can be recorded as a "background" timing signal on the same tape that is used to record observations, marking the tape with accurate timing signals.

It is useful to announce events during an occultation whether just audio or both audio and video recording is being used. The observer should record an appropriate comment for first contact, such as "off." Describe the disappearance, indicating if the object disappeared instantaneously, gradually, or in steps. If the "off" comment was late, then that should be noted along with an estimate of how late it was. This estimate is called a **personal equation**. Any difficulties in observing or unusual observations should also be described on the recording. When the object reappears, the observer should say "on," or something equivalent, and describe how the reappearance occurs, along with a personal equation, if necessary.

If accurate data from the occultation is desired, and an accurate timing signal was not recorded, it is important to determine if the recorder runs at the correct speed when recording and playing back. The position where the observing was done may also need to be accurately determined. This can be done using the Global Positioning Satellites. It can also usually be done within the territory of the United States by using quadrangle maps, which can be obtained from the US government printing office.

Procedures

Apparatus

telescopes, various eyepieces, short-wave radio capable of receiving and making audible timing signals from WWV or CHU, portable audio and/or video tape recorders, tapes and fresh or fully-charged batteries, maps giving locations of observing sites, star charts, and safe-lights.

A. Pre-observing Exercise (as a Class)

1. Review the predicted occultation to verify the date, time, location, the object's brightness, the phase and position of the Moon, position of the Sun (if it will interfere), and the position angles.

2. Since most publications listing occultation predictions are published to be read in many time zones, predicted times are usually given in Universal Time. Convert the predicted event times to the appropriate time zone times and date.

3. Carefully plan the observing event. Include a list of all the apparatus and steps which will be required at the observing site. If multiple observing stations are set up (as for grazing occultations, planetary occultations, and asteroidal occultations), carefully plan the locations of all observing stations.

4. Discuss plans to meet the personal needs of the participants. Plan the transport of both the personnel and equipment to the observing sites and back. Discuss the types of clothing that will be needed. Discuss arrangements for food and beverages, and insect repellents. Discuss the fact that toilet facilities may not exist at some remote observing sites.

5. Verify that each participant can operate the required equipment under the conditions expected at the observing site. This should include a "dress-rehearsal" of the observing exercise in the laboratory, noting if additional equipment or refined procedures are needed.

6. Describe what kind of report, if any, will be required from the students, and strategies for keeping observing notes.

B. Setting Up at the Observing Site

1. Help unpack and set up the telescope and other equipment at the observing site.

2. Tune in the short-wave receiver to WWV or CHU, make the timing signal audible, and run the tape recorders while describing the observing conditions. Play back some of the tapes to make certain all desired signals are being recorded.

C. Observing Lunar Occultations

1. Use the Moon as a reference and the position angle as a guide, locate the star or planet to be occulted. Use the position angle to locate the reappearance site on the Moon.

2. Follow the star or planet as it approaches the Moon. Make comments up to the disappearance (e.g., "The star is easy to see," "First contact will occur soon," and so on.) State "off" when the object dissapears. Verbalize your personal equation, if needed, and comment on any other observations or problems you experience.

3. Watch the area where the reappearance is expected, but remain alert to other events in the visual field. State "on" when the object reappears. Continue to record comments as the occultation ends, describing your personal equation, if needed.

4. Help pack up the equipment for its return to the laboratory.

D. Observing Grazing Occultations

1. Using the Moon as a reference, locate the star or planet to be occulted, and follow the object as it approaches the Moon. Make comments on tape about the ease of observing, etc.

2. Announce on tape first contact, second contact, etc. Describe your personal equation, if needed. Be prepared for a disappearance and reappearance at any time! Continue observing and commenting until it is clear that fourth contact has occurred.

3. Help pack up the equipment for its return to the laboratory.

E. Observing Planetary or Asteroidal Occultations

1. Find the star and planet or asteroid to be involved in the occultation, and follow the objects as they approach each other. Make comments on tape about their separation, their brightnesses, the observing conditions, etc.

2. Announce on tape the disappearance and reappearance. Describe your personal equation, if needed. Continue observing and commenting until it is clear the occultation is over.

3. Help pack up the equipment for its return to the laboratory.

F. Occultation Report

1. Describe the occultation event observed, the equipment and site used.

2. "Reduce" the recording(s) made while observing by replaying them and listing in a table the times when each important event occured. This may require replaying the tape several

times and also using a stop watch to measure time intervals between timing signals recorded on the tape and the time when observing events were recorded. If the data warrants, create a graphical display similar to that shown in Figure 30-3.

3. Evaluate and interpret your experience in observing occultations.

4. Describe your contributions in helping set up and pick up equipment at the observing site.

The Magnitude Scale

On a dark, clear night far from city lights, the unaided human eye can see on the order of five thousand stars. Some stars are bright, others are barely visible, and still others fall somewhere in between. A telescope reveals hundreds of thousands of stars that are too dim for the unaided eye to see. Most stars appear white to the unaided eye, whose cells for detecting color require more light. But the telescope reveals that stars come in a wide palette of colors. This lab explores the modern magnitude scale as a means of describing the brightness, the distance, and the color of a star.

The earliest recorded brightness scale was developed by Hipparchus, a natural philosopher of the second century BCE. He ranked stars into six magnitudes according to brightness. The brightest stars were first magnitude, the second brightest stars were second magnitude, and so on until the dimmest stars he could see, which were sixth magnitude. Modern measurements show that the difference between first and sixth magnitude represents a brightness ratio of 100. That is, a first magnitude star is about 100 times brighter than a sixth magnitude star. Thus, each magnitude is $100^{1/5}$ (or about 2.512) times brighter than the next larger, integral magnitude. Hipparchus' scale only allows integral magnitudes and does not allow for stars outside this range.

- With the invention of the telescope, it became obvious that a scale was needed to describe dimmer stars. Also, the scale should be able to describe brighter objects, such as some planets, the Moon, and the Sun. The modern magnitude scale allows larger magnitudes (for objects dimmer than sixth magnitude), zero and negative magnitudes (for objects brighter than first magnitude), and fractional magnitudes (for minor distinctions in brightness). As examples, the Sun appears at a magnitude of -26.7 in Earth's sky and the full Moon is 480 thousand times dimmer at magnitude -12.5.

The ratio of the brightness of one object (B_1) to the brightness of a second object (B_2) is given by

$$\frac{B_1}{B_2} = \left(100^{0.2}\right)^{(m_2 - m_1)}, \tag{1}$$

where m_1 and m_2 are the magnitudes of the first and second objects, respectively.

Example 1: How many times brighter is Sirius at magnitude -1.47 than the dimmest star visible to the unaided human eye (at about magnitude +6.0)?

$$\frac{B_{Sirius}}{B_{dim}} = \left(100^{0.2}\right)^{(+6.0-(-1.47))} = 970.$$

Thus, Sirius is almost a thousand times brighter than the dimmest star visible to the unaided human eye.

The magnitude of an object (m_2) can be determined by comparing its brightness with another object whose magnitude (m_1) is known. Let B_1/B_2 represent the brightness ratio of the known object to the unknown object. Then,

$$m_2 = m_1 + 2.5 \cdot \log_{10}\left(B_1 \Big/ B_2\right).$$ (2)

Example 2: An observer with a small telescope sees a comet that is 120 times dimmer than a star with magnitude +1.5. What is the magnitude of the newly seen comet?

The ratio of brightness of the brighter object to the dimmer one is 120 times.

Thus, the magnitude of the comet is
$$m_2 = +1.5 + 2.5 \cdot \log_{10}(120) = +1.5 + 2.5 \cdot 2.079 = +6.7.$$

Light-gathering power (LGP) is a measure of the amount of light collected by an optical instrument and is proportional to the area of the light-collecting surface. For most telescopes, this is proportional to the square of the diameter of the objective aperture (D). Mathematically, this is expressed as
$$LGP \propto D^2$$ (3)

Example 3: How many times more light is collected by a pair of 10X70. binoculars than by fully dark-adapted human eyes?

The diameter of the binoculars' objective aperture is 70. mm. Assume that the diameter of the fully-dilated pupil is 6.0 mm. Then,

$$\frac{LGP_{bin}}{LGP_{eye}} = \left(\frac{D_{bin}}{D_{eye}}\right)^2 = \left(\frac{70. \cdot mm}{6.0 \cdot mm}\right)^2 = 140$$

With 10X70. binoculars, one can see objects that are 140 times dimmer than the dimmest objects visible to the human eye. Another way to look at this is that the dimmest objects seen with the human eye are 140 times brighter than the dimmest objects seen through 10X70. binoculars.

Example 4: What is the dimmest magnitude visible to the dark-adapted, human eye (pupil diameter, 6.0 mm) through a 34 cm diameter telescope?

The telescope collects more light than the unaided eye. The LGP ratio of the telescope to the eye is

$$\frac{LGP_{tel}}{LGP_{eye}} = \left(\frac{340 \cdot mm}{6.0 \cdot mm}\right)^2 = 3200$$

Thus, the dimmest object visible to the unaided eye is 3200 times brighter than the dimmest object visible through the telescope.

The magnitude of the dimmest object visible through the telescope is
$$m_2 = +6.0 + 2.5 \cdot \log_{10}(3200) = +15.$$

A 34-cm diameter telescope can be used to view objects with magnitudes brighter (less) than +15.

The magnitude scale was used originally to distinguish stars according to how bright they appear from Earth. This is now called **apparent magnitude** (m). Table 1 lists the apparent magnitudes of some familiar objects.

Table 1: Some apparent magnitudes.

object or description	apparent magnitude
Sun	-26.7
Moon (full)	-12.5
Venus (brightest)	-4.6
Jupiter (brightest)	-2.5
Sirius (α Canis Majoris)	-1.5
Canopus (α Carinae)	-0.7
Vega (α Lyrae)	0.0
Unaided-eye limit	+6.0
Proxima Centauri	+11.1
Limit for 10x70 binoculars	+11.3
Limit for 15-cm telescopes	+13.0
Limit for 5-m telescope	
visual	+20.0
photographic	+23.5
electronic	+26.0
Limit for Hubble Space Telescope	+30.0

How bright an object seems from Earth depends on both the intrinsic brightness of the object and its distance from Earth. A star that appears dim may be a nearby, intrinsically dim star or a distant, intrinsically bright star. What is needed is a scale for describing the intrinsic brightness of an object. Once the intrinsic brightness of a star is known, its apparent magnitude for any distance can be computed. The apparent magnitude of a star at a distance of 10 parsecs is called the **absolute magnitude** (M) of the star, and is a standard way of expressing the intrinsic brightness of a star. The Sun, for example, has an apparent magnitude of -26.7, but its absolute magnitude is +4.85. If the Sun were 10 pc from Earth, it would be only 2.9 times brighter than the dimmest star visible to the unaided eye. Table 2 lists the apparent and absolute magnitudes of some stars.

Table 2: The apparent and absolute magnitudes of some stars and their color indices.

Name	apparent magnitude	absolute magnitude	Color Index (B - V)
Aldebaran (α Tau)	+0.86	-0.2	+1.54
Antares (α Sco)	+0.92	-4.5	+1.83
Bellatrix (γ Ori)	+1.64	-3.6	-0.22
Betelgeuse (α Ori)	+0.41	-5.5	+0.03
Canopus (α Car)	-0.72	-3.1	+0.15
Deneb (α Cyg)	+1.26	-6.9	+0.09
Fomalhaut (α PsA)	+1.19	+2.0	+0.09
Mimosa (β Cru)	+1.28	-4.6	-0.23
Pollux (β Gem)	+1.16	+0.8	+1.00
Regulus (α Leo)	+1.35	-0.6	-0.11
Sirius (α CMa)	-1.46	+1.4	0.00
Spica (α Vir)	+0.91	-3.6	-0.23
Sun (Sol)	-26.7	+4.85	+0.65
Wolf 359	+13.5	+16.7	+2.01

The **distance modulus** (m - M) is the difference between the apparent magnitude and the absolute magnitude of a star, and is a measure of the distance to the star. These quantities are related by

$$m - M = -5 + 5 \cdot \log\left(\frac{d}{1 \cdot pc}\right),$$ (4)

where d is the distance to the star measured in parsecs (pc). If the apparent and absolute magnitudes are known, the distance to the star can then be determined from

$$d = 10^{\frac{m-M+5}{5}} \cdot pc.$$ (5)

Example 5: Deneb (Alpha Cygni) has an apparent magnitude of +1.26 and lies at a distance of 430 parsecs. What is the absolute magnitude of Deneb? If Deneb were at the same distance as the Sun, how bright would it appear in Earth's sky?

$$M = m + 5 - 5 \cdot \log\left(\frac{430 \cdot pc}{1 \cdot pc}\right) = -6.9.$$

At a distance of 10 pc, Deneb has a magnitude of -6.9. At this distance, the Sun has a magnitude of +4.85.

$$\frac{B_{Deneb}}{B_{Sun}} = \left(100^{0.2}\right)^{(+4.85-(-6.9))} = 5.0 \cdot 10^4.$$

Hence, Deneb would appear 50. thousand times brighter than the Sun.

Example 6: Betelgeuse has an absolute magnitude of -5.5 and an apparent magnitude of +0.41. How far away is Betelgeuse in parsecs?

$$d = 10^{\frac{+0.41-(-5.5)+5}{5}} \cdot pc = 150 \cdot pc .$$

Apparent visual magnitude (V) of a star is its apparent magnitude at a wavelength of 550 nm, where the human eye is most sensitive. Early photographic film, however, was more sensitive to the blue region of the spectrum, with its greatest sensitivity around 450 nm. So, when photographs were taken of the sky, blue stars appeared brighter photographically than they did visually. Although, today, astronomers use photometers to obtain accurate measurements of a star's brightness at these wavelengths, the magnitude of a star based on its brightness at 450 nm is still called its **apparent photographic magnitude (B)**. See Exercise 12, Astrophotography, and Exercise 13, Electronic Imaging.

The **color index** of a star is the difference between its apparent photographic magnitude and its apparent visual magnitude. That is,

color index = B - V.   (6)

The color index is zero for white stars with surface temperatures around 9200 K. Since B is smaller (brighter) than V for a blue star, its color index is negative. Similarly, red stars have positive color indices. Our yellow-white Sun has a color index of +0.65 or, simply, 0.65. See Table 2 for more stars and their color indices.

Example 7: Bellatrix (γ Ori) has a color index of -0.22. How many times brighter does it appear at a wavelength of 450 nm than at a wavelength of 550 nm?

Here we wish to compare the brightness of magnitude B to that of V.
Using Equation 1,

$$\frac{B_B}{B_V} = \left(100^{0.2}\right)^{(V-B)} = \left(100^{0.2}\right)^{-(B-V)} = \left(100^{0.2}\right)^{-(-0.22)} = 1.2 .$$

Thus, Bellatrix is 1.2 time brighter at 450 nm than at 550 nm.

Procedures

Apparatus
list of current planetary magnitudes (available in most astronomy magazines).

A. Brightness
1. Rigel (Beta Orionis) has an apparent magnitude of +0.14 and Deneb (Alpha Cygni) has an apparent magnitude of +1.26. What is the ratio of the brightness of Rigel to Deneb?

2. How many times brighter is the full Moon (-12.5) than Venus at its brightest (-4.6)?

3. How many times dimmer is Venus currently than at its brightest?

4. How many times brighter is Jupiter than Mars, currently?

B. Light-Gathering Power
1. Determine the ratio of the LGP of a 25-cm diameter telescope to the dark-adapted, unaided human eye. Assume that the pupil is 6.0 mm in diameter.

2. What is the dimmest magnitude visible through a 25-cm diameter telescope when viewing with a fully dark-adapted human eye? The average human eye dilates to 6.0 mm when fully dark-adapted and can see to +6.0 magnitude.

3. What diameter telescope is needed to just barely see an object with an apparent magnitude of +18?

4. If all the planets were above the horizon, which would currently be visible to the dark-adapted, unaided human eye? Which would be visible if a pair of 7X35 binoculars were used?

C. Magnitude and Distance
1. How far away is Canopus (Alpha Carinae), which has an apparent magnitude of -0.72 and an absolute magnitude of -3.1?

2. What would be the magnitude of the Sun if it were at the same distance as Sirius (2.7 pc away)? The Sun's absolute magnitude is +4.85.

3. Make a table listing the distance moduli from 0 through 10 and the corresponding distances in parsecs.

D. Color Index

1. Of the stars in Table 2, list the ones more blue than the Sun.

2. Of the stars in Table 2, list the ones more red than the Sun.

3. Epsilon Indi has a color index of 1.05. What color would you expect it to appear to your eyes?

4. How many times brighter is Aldebaran (α Tau) at 550 nm than at 450 nm? Hint: here we are comparing V to B.

Hertzsprung-Russell Diagram

In the early 1900's two astronomers, Ejnar Hertzsprung and Henry Norris Russell, independently discovered a relationship between the luminosity of a star and its temperature. The Hertzsprung-Russell (HR) diagram is a plot of the luminosity (or absolute magnitude) against temperature (or spectral class) for a group of stars. See Preface, Spectral Classifications, for more information on spectral classes. As will be seen in this exercise, the HR diagram is one of the most powerful analytic tools available to astronomers. It can be used to estimate many characteristics of stars, such as size and mass.

Figure 32-1 shows a typical HR diagram. Spectral class or temperature is plotted along the horizontal axis from the highest temperatures on the left to the coolest temperatures on the right. Absolute magnitude is plotted along the vertical scale. **Luminosity** is the total amount of energy emitted by the star each second. A logarithmic, luminosity scale can also be used for the vertical axis. If luminosity is measured relative to the Sun's luminosity, the scale, for stars in our galaxy, ranges from 10^{-4} to 10^4.

Almost 90 percent of observed stars lie along a band that stretches from the upper left to the lower right on the HR diagram. This band is called the **main sequence**. Hot stars (left part of the diagram) tend to be bright (upper part) and cool stars (right part) tend to be dim (lower part). The photosphere temperatures of stars on the main sequence range from about 3000 K to about 30,000 K, a factor of 10. However, luminosities stretch over eight powers of ten.

It is now understood that all stars on the main sequence generate energy by the same mechanism, namely by fusing hydrogen into helium. The scattering of stars along the band of the main sequence is a consequence of their different initial masses.

For main sequence stars, there is a simple relationship between luminosity and mass. Stellar masses and luminosities increase from the lower right of the main sequence to the upper left. From this relationship, the mass of a main sequence star can be estimated by its position on the HR diagram. Unfortunately, there is no such simple relationship between luminosity and mass for the approximately 10 percent of stars that are not on the main sequence.

The upper right corner of the HR diagram is populated with cool but bright **giant stars**. These stars are brighter than main sequence stars of the same temperature. Since the luminosity of a star depends on both the temperature and the size of the star, red giants must be larger than main sequence stars of the same temperature. Small, hot stars whose low luminosities belie their high temperatures are called **white dwarfs**. White dwarfs fall to the left of the main sequence on the HR diagram. Masses vary widely in these regions off the main sequence, but typically, higher mass stars appear in the upper regions of the diagram.

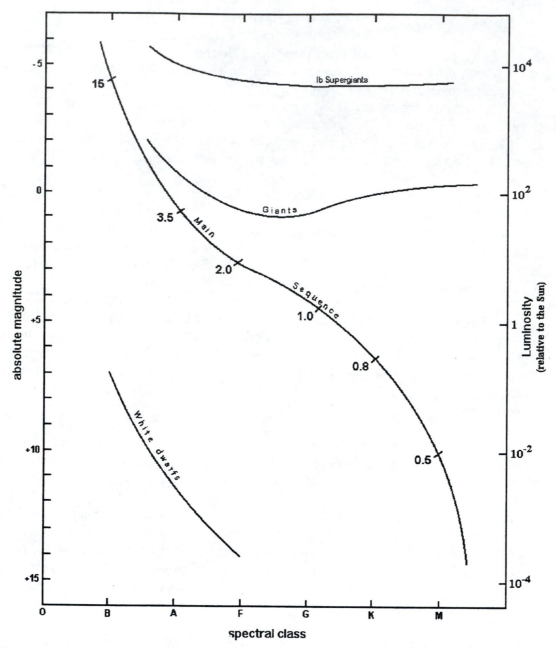

Figure 32-1: HR Diagram. Approximate stellar masses (relative to the Sun) are shown along the main sequence curve

The stars not on the main sequence have different mechanisms for generating energy. The giant stars fuse elements higher than hydrogen, which produces more power than the fusion of hydrogen. White dwarfs are the cores of Sun-like stars that have lost their outer layers. They no longer generate energy, but shine by radiating away energy stored in them from their previous stages.

Stars fuse lower elements into higher elements and thus, slowly change their compositions. This process is called **stellar evolution**. The rate of stellar evolution varies with the star's mass. A more massive star evolves faster than a less massive star because of its stronger gravity. Stronger gravity compresses the core more, increasing temperatures and pressures, which in turn increases the rate of fusion. The rate of energy output of stars on the main sequence increases nearly with the fourth power of the mass.

Stars with the mass of the Sun typically remain on the main sequence for 10 billion years. Stars with less mass than the Sun remain on the main sequence longer. Although lower mass stars have less hydrogen in their cores than the Sun, they also have lower core temperatures and pressures and consequently fuse their hydrogen more slowly. A 0.75 solar mass star remains on the main sequence for some 20 billion years, while a five solar mass star spends only about 70 million years on the main sequence. Stars much higher along the main sequence than the Sun must be younger than the Sun, else they would have used up the hydrogen in their cores by now and would have moved off the main sequence.

Stars form from huge, cool nebulae. As a nebula collapses under the attraction of gravity, it fragments into lumps, called **protostars**. Gravitational energy from the collapse of the protostar is converted to heat and raises the temperature of the protostar. When the temperature reaches about 3000 K, the "photosphere" of the protostar emits red light. When the early Sun reached this point, its radius was 20 times larger than it is currently, and, thus, it appeared nearly 30 times brighter. This placed it just to the right of the main sequence and above its current position on the HR diagram.

The various stages of a star's "life" occupy different points on the HR diagram. As the star passes through these stages, it follows a path on the HR diagram. A plot of a star's evolution on an HR diagram is called its **evolutionary track**. Astronomers can use the position of a star on the HR diagram to estimate its age.

The early Sun continued to collapse under gravity until its core reached a temperature of about 7 million kelvins and a pressure of some 10 billion atmospheres. Under these conditions, hydrogen began to fuse into helium, and the Sun became a star. The Sun's evolutionary track had reached the main sequence. A one solar mass protostar typically takes about 30 million years to reach the main sequence.

If a protostar's mass is less than about 0.08 solar masses, it never has a high enough core temperature or pressure to begin hydrogen fusing. Such a protostar, called a **brown dwarf**, would slowly contract, dimly radiating away gravitational energy as mostly infrared light. If the core temperature of the brown dwarf reaches about a million kelvins, it fuses its supply of deuterium to helium. Since the deuterium is not replaced, this fusion eventually stops, and the brown dwarf resumes its contraction and cooling to eventually become an extremely large planet. Stars with masses greater than about 80 times that of the Sun tend to develop such high temperatures that radiation pressure from rapid fusion quickly tears them apart. Thus, main sequence stars may have masses only between about 0.08 and 80 solar masses.

The Sun is expected to stay on the main sequence for some 10 billion years, until it has fused most of the hydrogen in its core to helium. The Sun has been on the main sequence for about 4.6 billion years, so, in about 5 billion years, hydrogen fusing will cease and the core will no longer be able to resist collapsing from its own weight.

As the core collapses, its temperature and density will increase. When the core's temperature reaches about 100 million kelvins, helium will begin to fuse rapidly into carbon and oxygen in what is called the **helium flash**. Fusion of helium produces more energy than fusion of hydrogen, so the temperatures in layers surrounding the Sun's core can reach the 7 million kelvins required for hydrogen fusing. Shells of hydrogen fusing will develop around the core, increasing the internal temperature of the Sun and causing the outer layers to expand. The photosphere of the Sun may swell past the orbit of Earth and cool to a reddish color at about 3500 kelvins. At this point, the Sun is said to be in its **red giant stage**. It will have moved off the main sequence upward and to the right.

The Sun may fluctuate in size as it adjusts to its new sources of energy. As energy production stabilizes, the Sun may settle into a region on the HR diagram about halfway between its main sequence position and its red giant stage. This region of the HR diagram is sometimes called the **clump**, as Sun-like stars may remain there for a couple billion years.

Helium fusion runs at a faster rate than hydrogen fusion. When most of the core's helium has been fused into carbon and oxygen, helium fusing in the core will shut down and the core will begin to collapse again. As the core's temperature and density increase, helium fusion will begin in shells nearest the core and the hydrogen-fusing shells will move outward. The outer layers will swell outward even further than in the red giant stage. The Sun's evolutionary track will leave the clump and moves up to the **red supergiant region** along a path, called **the asymptotic giant branch**. Stars in this stage exhibit absorption spectra of neutral atoms and molecules, as their photospheres have cooled to about 2000 K and electron bonds can remain stable.

During the ascent of the asymptotic giant branch, other nuclear reactions will occur in the Sun. For example, slow neutron capture will produce higher elements up through bismuth (with 83 protons). The Sun will experience deep convection currents, too, which distribute the higher elements throughout the Sun, even to the outermost layers.

The Sun may pulsate with increasingly violent expansions and contractions as multiple helium flashes occur in the shells surrounding the core. The evolutionary track will move horizontally to the left through a region of the HR diagram called the **instability strip**.

A powerful solar wind will develop that will spew out about 10^{-5} solar masses of material each year. This will create a shroud of gas and dust around the Sun. About half of the Sun's mass may be lost in this way, leaving behind the hot, naked core. The Sun will then be called a white dwarf.

The core will continue to shrink until it reaches about the size of Earth. Its temperature may exceed 100 thousand kelvins. The electrons of the core will be packed so tightly that the Pauli exclusion principle will prevent further contraction. The **Pauli exclusion principle** was defined in 1925 by the German physicist, Wolfgang Pauli, and sets a limit on how tightly matter containing electrons can be packed. The core material of a white dwarf is packed to about 10^7 times denser than liquid water. So, a liter of the core material has a mass of 10 thousand tons.

The Sun, as a hot white dwarf, will emit a great deal of ultraviolet light, which will heat and ionize the surrounding shroud of gas and causes it to fluoresce. Such a nebula is called a **planetary nebula** from its appearance in early telescopes, though it has no relationship to planets. The planetary nebula will continue to expand, cool, and dim. After a million years, the nebula may no longer be visible. However, it may take many billions of years for the Sun to cool to a dark cinder, called a **black dwarf**.

Procedures

Apparatus

Photocopies of the HR diagram, and the Temperature and Radius overlays on transparency films, if possible.

A. Plotting the Brightest Stars Seen from Earth

1. Plot the stars listed in Table 1 (the brightest stars seen from Earth) on the HR diagram (Figure 32-1) by marking a dot on the diagram corresponding to each star's absolute magnitude and spectral class.

2. What percentage of these stars are main sequence stars?

3. What percentage of these main sequence stars are more massive than the Sun?

4. Approximate the temperatures of the stars in Table 1 by overlaying your HR diagram with the Temperature Overlay (Figure 32-2). Make a table containing these stars' names and approximate temperatures. Arrange the table in order of increasing temperature.

5. What percentage of the stars in Table 1 are hotter than the Sun?

6. Approximate the sizes of the stars in Table 1 by overlaying your HR diagram with the Radius Overlay (Figure 32-3). Arrange these values in the table created in step A.4.

7. What percentage of the stars in Table 1 are larger than the Sun?

B. Plotting the Nearest Stars to Earth

1. Plot the stars listed in Table 2 (the nearest stars to Earth) on the HR diagram. Plot these stars in a different color on the same HR diagram used in section A.

2. What percentage of these stars are main sequence?

3. What percentage of these main sequence stars are more massive than the Sun?

4. Approximate the temperatures of the stars in Table 2 by overlaying your HR diagram with the Temperature Overlay (Figure 32-2). Make a table containing these stars' names and approximate temperatures. Arrange the table in order of increasing temperature.

5. What percentage of the stars in Table 2 are hotter than the Sun?

6. Approximate the sizes of the stars in Table 2 by overlaying your HR diagram with the Radius Overlay (Figure 32-3). Arrange these values in the table created in step B.4.

7. What percentage of the stars in Table 2 are larger than the Sun?

Table 1: Some of the Brightest Stars Seen from Earth

Name	Apparent Magnitude	Absolute Magnitude	Spectral Class
Sirius A (Alpha Canis Majoris A)	-1.46	+1.4	A1V
Canopus (Alpha Carinae)	-0.72	-3.1	F0II
Rigel Kentaurus A (Alpha Centauri A)	-0.01	+4.4	G2V
Arcturus (Alpha Boötis)	-0.06	-0.3	K2IIIp
Vega (Alpha Lyrae)	+0.04	+0.5	A0V
Capella A (Alpha Aurigae A)	+0.05	-0.6	G8III
Rigel A (Beta Orionis A)	+0.14	-7.1	B8Ia
Procyon A (Alpha Canis Minoris A)	+0.37	+2.7	F5IV
Betelgeuse (Alpha Orionis)	+0.41	-5.6	M2Ia
Achernar (Alpha Eridani)	+0.51	-1.0	B3V
Hadar A (Beta Centauri A)	+0.63	-5.2	B1III
Altair (Alpha Aquilae)	+0.77	+2.2	A7IV
Acrux A (Alpha Crucis A)	+1.39	-3.9	B1IV
Aldebaran A (Alpha Tauri A)	+0.86	-0.7	K5III
Spica (Alpha Virginis)	+0.91	-3.6	B1V
Antares A (Alpha Scorpii A)	+0.92	-4.5	M1Ib
Pollux (Beta Geminorum)	+1.16	+0.8	K0III
Fomalhaut A (Alpha Piscis Austrini A)	+1.19	+2.0	A3V
Deneb (Alpha Cygni)	+1.26	-6.9	A2Ia
Beta Crucis A	+1.28	-4.6	B0.5IV

Table 2: Some of the Nearest Stars

Name	Apparent Magnitude	Absolute Magnitude	Spectral Type
Sun	-26.72	+4.85	G2V
Proxima Centauri	+11.05	+15.4	M5e
Rigel Kentaurus	-0.01	+4.4	G2V
Barnard's Star	+9.54	+13.2	M5V
Wolf 359	+13.53	+16.7	M8V
BD +36°2147	+7.50	+10.5	M2V
Luyten 726-8	+12.52	+15.3	M5.5V
Sirius	-1.46	+1.4	A1V
Ross 154	+10.45	+13.3	M4.5V
Ross 248	+12.29	+14.8	M6V
Epsilon Eridani	+3.73	+6.1	K2V
Ross 128	+11.10	+13.5	M5V
61 Cygni	+5.22	+7.6	K5V
Epsilon Indi	+4.68	+7.0	K5V
BD +43°44	+8.08	+10.3	M1V
Luyten 789-6	+12.18	+14.6	M6V
Procyon A	+0.37	+2.6	F5IV
BD +59°1915 A	+8.90	+11.2	M3V
CD -36°15693	+7.35	+9.58	M1.3eV
Tau Ceti	+3.5	+5.72	G8V

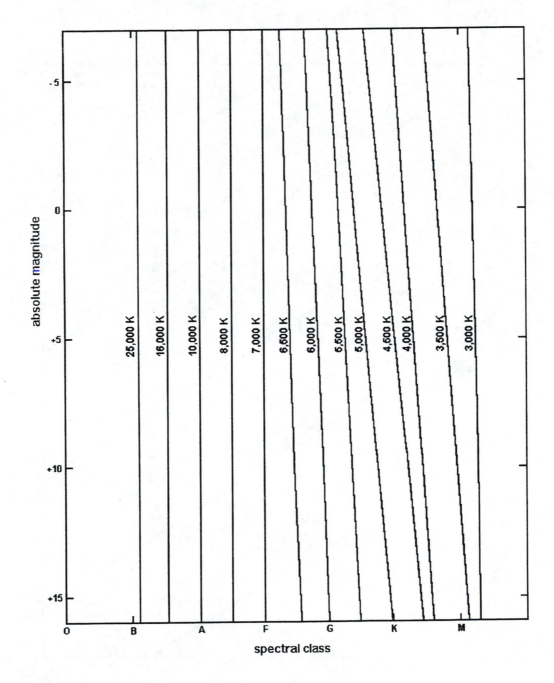

Figure 32-2: Temperature Overlay.

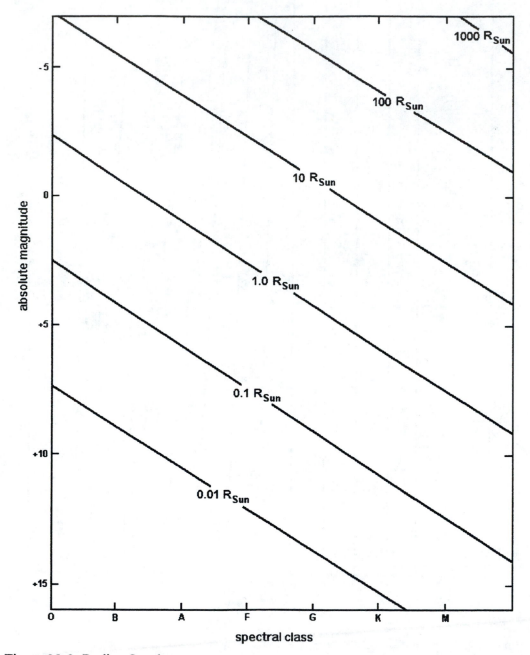

Figure 32-3: Radius Overlay.

Elements and Supernovae

Stars and nebulae make up the luminous matter of galaxies and are the subject of much of astronomy. They are mostly made of atoms, including atoms in the forms of ions and molecules. The Bohr model of the atom is described in Exercise 11, Atomic Spectra. **Ions** are atoms which have lost or gained one or more electrons. Atoms in the Sun, for example, where the temperature is 5800 K and up, are ionized. At these temperatures, atoms move with such high speeds that electrons are often knocked free when they collide. This exercise explores basic ideas about atoms, including where they are made, how long they may last, and how they are being distributed in the Universe. Also see Exercise 42, Radioactivity and Time.

Most matter encountered on Earth is made of atoms and molecules. **Molecules** are clusters of atoms bound together in specific arrangements by electromagnetic forces. Molecules of oxygen, for example, consist of two oxygen atoms. Other familiar examples of molecules are molecular nitrogen (two bound nitrogen atoms), water (a cluster of three atoms, one oxygen and two hydrogen atoms), ammonia (one nitrogen and three hydrogen atoms), methane (one carbon and four hydrogen atoms), and sugar (an arrangement of 6 carbon, 12 hydrogen, and 6 oxygen atoms). Matter not made of atoms includes electron beams, neutrons, and other examples given below.

Each atom is identifiable as being an **isotope** of a particular element. All atoms of an element have the same number of protons in their nuclei, and also the same number of electrons when the atoms are not ionized. Different isotopes of an element have nearly identical chemical properties, but have different masses and nuclear properties, as described below. Periodic tables list some properties of each element, as can be seen in the periodic table of Figure 33-1. Studying atoms is appropriate in astronomy since their properties help determine the forms and natures of many celestial bodies. Also, the types of atoms present in an object provide clues to its history.

Around 400 BCE, the Greek philosopher Democritus theorized the existence of atoms. But a detailed theory was not provided until 1803 when John Dalton, an English chemist, expressed the idea that a limited number of atomic types form all of the molecules and compounds that make up the wide variety of materials found on Earth. In 1869, Dmitri Mendeleev, a Russian scientist, published the first periodic table, which placed the different elements then known in rows of increasing mass, and those elements with similar chemical properties in the same columns. This table contained empty spaces, which suggested additional elements were yet to be discovered. The properties of undiscovered elements could be predicted from their positions in the table. The elements gallium, scandium, and germanium were soon discovered using this information.

The chemical properties of an atom are determined by the number of electrons in the atom's electrically neutral form. The electrons occupy the outer regions of the atom, and are equal in number to the protons in the nucleus. Current periodic tables, like the one in Figure

PERIODIC TABLE OF THE ELEMENTS

Table of Selected Radioactive Isotopes

Periodic table chart of the elements showing atomic numbers, atomic weights, symbols, electron configurations, boiling points, melting points, densities, and oxidation states for each element. Includes the main table, the lanthanide and actinide series, a table of selected radioactive isotopes, and a KEY diagram explaining the data layout.

KEY:
- ATOMIC NUMBER
- ATOMIC WEIGHT (2)
- OXIDATION STATES (Bold most stable)
- SYMBOL (1)
- ELECTRON CONFIGURATION
- NAME
- BOILING POINT, K
- MELTING POINT, K
- DENSITY at 300K (3) (g/cm³)

Example shown: 30, 65.38, Zn, 1180, 692.73, 7.14, [Ar]3d¹⁰4s², Zinc

NOTES:
(1) Black — solid. Red — gas. Blue — liquid. Outline — synthetically prepared.

(2) Based upon carbon-12. ( ) indicates most stable or best known isotope.

(3) Entries marked with daggers refer to the gaseous state at 273 K and 1 atm and are given in units of g/l.

* Estimated Values

SARGENT-WELCH
VWR SCIENTIFIC

911 Commerce Court • Buffalo Grove • Illinois 60089
(800) 727-4368 • Fax (800) 676-2540

Catalog Number S-18806-10

Side 1

The A & B subgroup designations, applicable to elements in rows 4, 5, 6 and 7, are those recommended by the International Union of Pure and Applied Chemistry.

The names for elements 104-106 have been proposed, but not formally accepted by the IUPAC.

Selected Radioactive Isotopes

Naturally occurring radioactive isotopes are designated by a mass number in blue (although some are also manufactured.) Letter m indicates an isomer of an element. Isotopes of the same mass number. Half-lives follow in parentheses, where s, h, d, and y stand, respectively for seconds, minutes, hours, days, and years. The table includes mainly the longer-lived radioactive isotopes; many others have been prepared. Isotopes known to be radioactive but with half-lives of a second or greater are listed. Symbol designating the principal mode (or modes) of decay (these processes are generally accompanied by gamma radiation):

- α alpha particle emission
- β⁻ beta particle (electron) emission
- β⁺ positron emission
- EC orbital electron capture
- IT isomeric transition from upper to lower isomeric state
- SF spontaneous fission

1, identify each element based on the number of protons in the atom's nucleus, from 1 for hydrogen through 92 for uranium. Additionally, some trans-uranium elements, which have been created in nuclear laboratories, are also included.

Each element has a number of isotopes; for example, hydrogen exists as hydrogen 1, hydrogen 2 (called deuterium), and hydrogen 3 (called tritium). Deuterium has a neutron in the nucleus along with its proton, while hydrogen 1 only has a proton in its nucleus. Tritium has two neutrons in its nucleus. The **relative abundances** of these isotopes on Earth are 99.985% hydrogen 1 and 0.015% hydrogen 2. There are only trace amounts of hydrogen 3.

Each tritium atom is **radioactive** and has a probability of decaying each second. With a large number of tritium atoms, the time required for half of the atoms to decay is 12.3 years, which is called tritium's **half-life**. Each radioactive isotope has a unique half-life. Each atom of tritium that decays emits an electron and a neutrino, leaving an atom of the isotope helium 3. Helium 3 has two protons and one neutron in its nucleus.

Hydrogen is said to be a **stable** element because it has two non-radioactive isotopes, hydrogen 1 and hydrogen 2. All elements have radioacitve isotopes, although for most elements, the radioactive isotopes are not present on Earth. Helium, for example, has two stable isotopes, helium 3 and helium 4. When Helium 6 is made it decays with a half-life of less than one second, so after a few minutes none of the helium 6 remains.

All but two of the elements from hydrogen to uranium occur naturally on Earth. The two "missing" elements are technetium, with 43 protons, and promethium, with 61 protons. Both have only radioactive isotopes. Most of the elements from hydrogen to uranium, some 81 of the 92, have stable or non-radioactive isotopes. One may say that stable isotopes have infinite half-lives. Atoms of stable isotopes should last as long as the Universe does, provided they are not put into some sort of nuclear reaction, such as in a nuclear laboratory or in the core of a star. The isotopes of the "missing" elements with the longest half-lives are, respectively, technetium 98 at 4.2 million years and promethium 145 at 18 years. Other isotopes of these elements have shorter half-lives and are said to be less stable.

No elements above bismuth, which has 83 protons, are known to have any stable isotopes, though isotopes of thorium and uranium have very long half-lives. All of the elements from bismuth to uranium are created by the radioactive decay of thorium, uranium, and their radioactive descendants, as shown in Figure 1 of Exercise 42, Radioactivity and Time. An isotope of element 111 was created in 1995 in a nuclear laboratory in Germany, where it decayed with a half-life of about 1 millisecond. It is uncertain whether stable isotopes of the trans-uranium elements can be found.

A basic question about elements is "Are the types of elements found on Earth common in the Universe?" Spectroscopic studies of x-ray, ultraviolet, visible, infrared, and radio emanations from extraterrestrial sources demonstrate that the elements found here are widely distributed in the Universe, where they produce identical radiation. The Fraunhofer lines in the solar spectrum, for example, allow 67 elements to be identified in the chromosphere of the Sun, as they produce spectra identical to spectra produced by the respective elements on Earth. See Exercise 11, Atomic Spectra.

The right-hand column of the periodic table lists the elements called inert gases. These elements are helium, neon, argon, krypton, xenon, and radon. These elements usually do not react chemically with other elements or themselves, and consequently have low melting and boiling point temperatures. As a result, these elements exist as gases in normal Earth conditions. If only these elements and hydrogen existed in the Solar System, there would be no inner planets where the terrestrial planets, Mercury, Venus, Earth, and Mars are now, as these gases would only condense to form solids further from the Sun where is it colder. Many of the metals, on the other hand, have high melting and boiling point temperatures, and consequently they condense to form liquids and solids in the inner Solar System.

All elements to the left of the bold line running just below boron, arsenic, and astatine on the periodic table are metals. Note that they make up the majority of the elements. Spectral lines from hydrogen and helium are found in the spectra of all stars, illustrating universal distribution of those atoms. If the spectrum of a star also shows many metals, the star is said to be a **metal-rich star**. Metal-rich stars populate the galaxy disk, and are also referred to as **galaxy disk stars** and **population I stars**.

The spectra from the oldest stars show almost no lines for elements other than hydrogen and helium. These stars are identified as **metal-poor stars** or **population II stars**. Since these population II stars are mostly found in globular clusters and the galactic halo, they are sometimes referred to as **globular cluster stars** and **galaxy halo stars**.

The observations on the metal content of stars are consistent with the idea that early in the history of the Universe only hydrogen, helium, and a trace of lithium were present, and that stars which formed then were metal-poor stars. Some of these stars still exist. Stars generate energy by fusing elements into higher elements in thermonuclear processes in their hot cores. Some of these higher elements are redistributed back into space when the stars explode as supernovae. The supernova processes, by fusion and also by adding neutrons to elements, generate all the elements in the periodic table, including many radioactive isotopes. These materials are blown out of the supernovae and mix with the dust and gas between stars. New population I stars, perhaps with orbiting asteroids and terrestrial planets, are believed to form from these enriched nebulae.

It is a common experience that when dropping an object, it never bounces as high as from where it fell. Just before exploding, the core of a supernova collapses and the outer layers fall on top of each other towards the core. How then do the layers of material falling into a supernova bounce completely out of the star? This process will be modeled below by dropping balls on top of each other.

Hydrogen, helium, and traces of lithium were created in the early Universe and formed nebulae and stars. Massive stars fuse higher elements in their cores, reaching iron, with 26 protons. In this stage, stars become unstable and explode as supernovae, a process in which all elements are both created and thrown out of these stars. This material includes many radioactive elements, which are observed to decay, producing characteristic radiation and also heating the expanding supernova nebula. Stellar systems forming from such nebulae would, of course, contain radioactive materials. These stellar systems would differentiate based on the temperature variations caused mainly by the heat generated by the central star. Near the forming star, only materials that have high melting and boiling temperatures could form solids and collect together to form planets. These would be terrestrial planets, containing mostly metals, metal oxides, and silicates (rocks). Further out, where it is colder, materials that have lower melting and boiling temperatures would also condense to form comets and planets, which would containing mostly ices and gases. Figure 33-2 shows the temperature versus distance from the Sun which is believed to have resulted as the planets were forming.

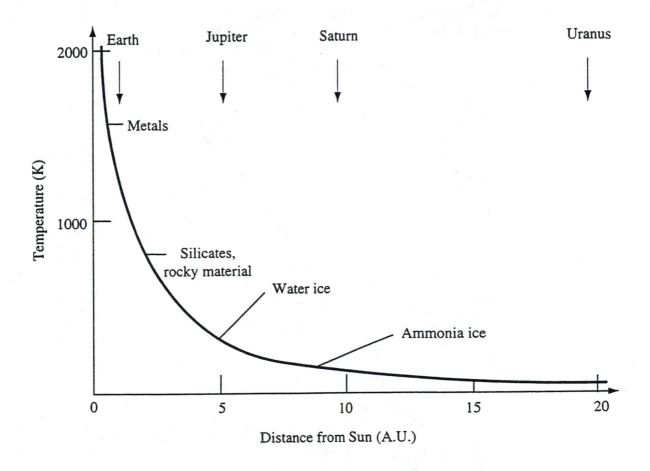

Figure 33-2. A plot of temperature versus distance from the primitive Sun.

Radioactive isotopes in the Earth have come from three different sources: long-lived isotopes (with half-lives on the order of a billion years or more, which have survived as radioactive atoms since this material was created), radioactive isotopes that are descendants of long lived isotopes, and radioactive isotopes otherwise created after Earth formed. The nuclear power and weapons industries, for example, have created a wide variety of radioactive isotopes, some of which have been released into the atmosphere. Carbon provides an informative example. Three isotopes of carbon exist on Earth, carbon 12 and carbon 13, which are stable, and carbon 14, which is radioactive with a 5730 year half-life. Carbon 14 is created naturally in the upper atmosphere by cosmic rays striking nitrogen 14 atoms there. Carbon 14 was also created by atomic weapons testing, which nearly doubled the amount of carbon 14 in the atmosphere.

The idea that everything we normally encounter can be described as being made of atoms and molecules is such a powerful idea, that some people go on to make the incorrect assumption that all matter is composed of the elements found in the periodic table. Types of matter not composed of elements include neutron stars, black holes, possible dark mass, and sub-atomic particles such as electrons, neutrinos, and neutrons. **Dark mass** or **dark matter**, a hypothetical material required to provide the gravity needed to hold galaxies together, has not been detected by any method other than its gravitational influence on visible matter. There is speculation that since dark mass has not been detected with traditional methods, it does not interact electromagnetically. That is, it will not absorb, emit, reflect, or scatter light of any wavelength. There are indications that the Universe may contain more dark mass than anything else. If this is true, then not only do atoms and elements not constitute all of the mass of the Universe, they do not even constitute a majority of the mass. Nevertheless, the elements have provided a rich collection of objects for astronomers and other scientists to study.

Procedures

Apparatus

Three balls of increasing mass, say a ping-pong ball, a practice (hollow hard plastic) golf ball, and a regular (regulation) golf ball, and 1 meter of string.

A. Elements and the Periodic Table

1. Look over and become familiar with the periodic table. Record in your report the chemical symbols used for the following elements: hydrogen, helium, oxygen, iron, mercury, and technetium. Note which elements are gases, solids, and liquids at ambient conditions for the surface of Earth (300 K and 1 atmosphere of pressure) by recording the melting and boiling point temperatures of these elements and noting if these temperatures are above or below the ambient temperature.

2. List the chemical symbols, names, and melting and boiling points for the inert gases. Which would form ices at 100 K, a characteristic temperature of the outer Solar System?

3. Note the outlined symbol's for technetium, promethium, and curium. The longest lived isotopes of these elements are given in the Table of Selected Radioactive Isotopes on the periodic table. List sequentially all isotopes and half-lives given for these three elements.

4. The abundance of radioactive isotopes decreases by half each half-life. So, after a number of years, say t, the abundance will have decreased by $(1/2)^{(t/T)}$, where T is the isotope's half-life. Construct a table giving the value of this function when (t/T) has the values of 2, 4, 8, 16, 32, and 64. Note this can be done by squaring (1/2) once, twice, thrice, four times, five times, and six times. (On scientific calculators, enter 0.5 and depress the "x^2" key once for the 2nd power, again for the 4th power, etc. As a test, the 64th power of 0.5 should be about 5×10^{-20}.)

Then, use the table created to find about how many years it will take for the longest lived isotopes of technetium and promethium to have decreased by at least a billion times (which occurs in 30 half-lives). Are these values consistent with these elements not being found on a planet believed to be 4.6 billion years old?

5. Notice that no melting or boiling point data are given for the five highest elements. Why might this be so? Justify your answer by reference to the half-lives of these elements.

6. Record in your report all of the radioactive isotopes of carbon and their half-lives.

7. Use melting and boiling point values to explain why meteoroids in the inner Solar System may be expected to contain bulk iron, nickel, cobalt, and carbon, but not bulk hydrogen, helium, nitrogen, or oxygen.

8. Use melting and boiling point values to explain why comets in the outer Solar System, say in the Kuiper belt beyond the orbit of Neptune, may contain bulk argon, hydrogen,

nitrogen, and oxygen. Note that ammonia, carbon dioxide, methane, water, and other compounds, which are important in comets and astronomy, are not elements, and so are not included in periodic tables of elements, or in this exercise.

9. Identify and record two elements with densities above 20 g/cm³. Note the density of water, which is 1 g/cm³, is used as a reference. Why would differentiation of a planet make bulk platinum and gold rare on the planet's surface? Would these materials likely be as rare near the surface of an undifferentiated asteroid?

10. Identify and record the elements with the highest melting and boiling point temperatures you can find. List those temperatures.

11. Identify and record the elements with the lowest melting and boiling point temperatures you can find. List those temperatures.

12. Arrange, in order from longest to shortest, the half-lives of the isotopes of uranium and thorium given in the table.

B. Supernova Core Bounce Model

1. Label the balls with the numbers 2, 3, and 4, in order of decreasing mass. The Earth will be ball number 1.

2. Stand a meter stick vertically on a hard surface, such as a table top or hard floor, and drop ball 4 from a position where the bottom of the ball is one-half meter above the surface. Observe and record how high the bottom of this ball bounces. Repeat if nessecary for ball 4, and then for balls 3 and 2. Make a table of your data giving the average return height. Note that energy lost to friction and in making sound should prevent the ball from returning to the original height.

3. Make a string hanger for the balls of two pieces of string about one-half meter long by placing the four string ends together and pulling the strings into two loops. Tie the four ends together into a knot so the two loops are exactly the same length.

Hold the loops by placing the knot between your teeth, then hang ball 2 in the loops. When in position, the ends of the loops opposite the knot should cross under the bottom of ball 2 at right angles, as shown in Figure 33-3. With ball 2 in place, position ball 3 on top of ball 2, and then ball 4 on top of ball 3, as shown.

4. Hold the string and three balls beside a meter stick held vertically over the same hard surface so the bottom of ball 4 is half a meter above the surface. Make sure the balls are not swinging, and release the string and balls. Repeat this several times, estimating how high each ball bounces. Ball 4 should have bounced much higher than from where it was dropped. Where did the energy come from to make it bounce that high? Hint, how high did balls 2 and 3 bounce?

5. Briefly explain how these bouncing balls model part of a supernova explosion.

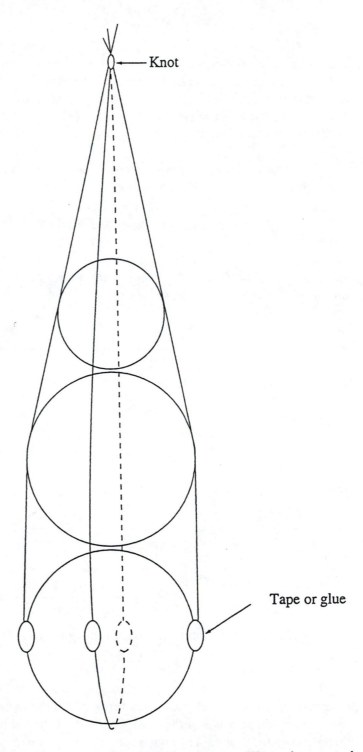

Figure 33-3. Illustrating the three balls suspended in the string hanger. The strings may be taped or glued around the sides of ball 2, as indicated.

Blackbody Radiation

A **blackbody** is an ideal object that completely absorbs any radiation that falls on it and emits radiation in a way that depends only on its temperature. All blackbodies of the same temperature emit the same characteristic radiation, known as **blackbody radiation**. Stars approximate blackbodies, so laws that describe the radiation emitted by blackbodies also describe the radiation emitted by stars. This lab deals with a few of the blackbody radiation laws and how they relate to some stellar characteristics.

A plot of the light intensity versus wavelength (or frequency) of the light emitted by a blackbody is called a **blackbody curve**. This curve has the shape of a lopsided bell curve with a steeper slope on the shorter wavelength side. The peak in the curve corresponds to the wavelength of the "brightest" color produced by the blackbody. Here, "brightness" refers to the highest intensity light and is irrespective of the perception of the human eye. Figure 34-1 shows the blackbody curves for an object at different temperatures. The Sun approximates a blackbody at a temperature of about 6000 K. The peak wavelength produced by the Sun is about 480 nm.

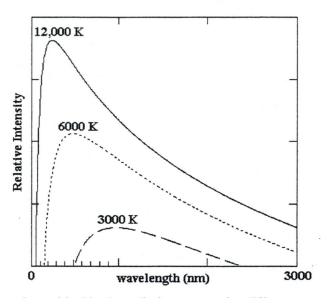

Figure 34-1: Some blackbody radiation curves for different temperatures.

When an object is slowly heated, the color of the light emitted by the object changes. Consider a piece of metal placed in a furnace. As the metal warms, it begins to glow a dull red. As the temperature of the metal continues to rise, it becomes more orange in color, then yellow, and white. Through a spectrometer, one sees that the metal emits mostly red light at first, and that colors of progressively shorter wavelengths are added as the metal warms. All the colors of the rainbow (red, orange, yellow, green, blue, and violet) are visible when the metal is white-hot.

This relation between temperature and color is stated in **Wien's law**: the wavelength at the peak of the blackbody curve (λ_{max}) is inversely proportional to the temperature of the object (T), or mathematically,

$$\lambda_{max} \propto \frac{1}{T}. \tag{1}$$

This relation gives a method for determining the temperature of a star.

Example 1: The brightest color produced by an object at a temperature of 6000. K is at a wavelength of 480 nm (blue). What is the wavelength of the brightest color produced by a 3000. K object?

$$\lambda_{max2} = \lambda_{max1} \cdot \left(\frac{T_1}{T_2}\right) = \left(480 \cdot nm\right)\left(\frac{6000. \cdot K}{3000. \cdot K}\right) = 960 \cdot nm \; .$$

The 3000. K object emits most of its light in the infrared portion of the spectrum. Since the blackbody curve is roughly bell-shaped, some visible light from the red end of the spectrum is also emitted. Hence, this object would appear red to human eyes, whereas a 6000. K blackbody would appear yellow-white.

Example 2: The brightest color emitted by a certain star has a wavelength of 240 nm. Given that a 6000. K star produces its peak radiation at a wavelength of 480 nm, what is the temperature of this star?

$$T_2 = T_1 \cdot \left(\frac{\lambda_{max1}}{\lambda_{max2}}\right) = \left(6000. \cdot K\right)\left(\frac{480 \cdot nm}{240 \cdot nm}\right) = 12000 \cdot K$$

In science and engineering, **power** is defined as the time rate at which energy is emitted or transferred. In metric units, power is measured in watts (W), where one watt is one joule of energy (or work) transmitted per second. A common, 100 W incandescent bulb requires 100 joules of energy every second to heat its tungsten filament.

A common incandescent lamp consists of a coiled filament of tungsten enclosed in a bulb of glass from which most of the air has been removed. The melting point of tungsten is 3655 K, so lamps using a filament made of tungsten must operate significantly below this temperature. Typically, the filament temperature is 2800 K, which produces a noticeably whiter light than a candle or kerosene lamp. Candles and kerosene lamps produce light by burning their hydrocarbon fuels, heating carbon particles in the flame to incandescence at about 2000 K. Gas discharge lamps typically have effective temperatures greater than 3000 K.

A body can lose heat by three mechanisms: radiation, convection, and conduction. Isolated objects in space lose heat only by emitting electromagnetic radiation (light). Objects usually encountered on Earth's surface are much too cool to emit visible light, but they emit infrared light, which may be imaged with an infrared camera. Convection requires an object to be immersed in a fluid medium, such as air or water. Then the medium may carry away heat by its motion. Conduction occurs when heat is transferred from a warmer object to a cooler one through contact.

Earth provides many examples of these mechanisms of heat transfer. Earth's surface is heated by light from the Sun, i.e. by radiation. The ground conducts this heat to some depth and across the surface. The ground warms air near it, making the air expand and become less dense. This air is now more buoyant than the cooler, denser air above it, so it rises and the cooler air takes its place. Convection currents are established as the warm air cools and falls back to the ground where it is heated and rises again. The warmed surface radiates some of its heat away as infrared light. Some of this infrared radiation is absorbed by gases in the atmosphere, such as carbon dioxide and methane, and some is radiated back into space. This partial trapping of the heat from the Sun is called the **greenhouse effect**. The current equilibrium between the incoming heat from the Sun and the heat lost to space keeps Earth at an average temperature of about 300 K.

The adult human body produces about the same power, averaged over a day, as a 100 watt lamp. The power output is greater when the body is active and less when the body is resting. The glass bulb of the common 100 watt incandescent lamp typically reaches a temperature of 500 K. Since the same power is produced by both the human body and the 100 watt lamp, why does the glass bulb become so much hotter? The answer comes from the size of the objects and how the heat is dissipated. The human body has a much larger surface area than the glass bulb, so the heat is distributed over more area where it can be radiated away. Also, the human body can dissipate heat by breathing out warmed air and by evaporating water from its surface. Both the human body and the glass bulb of the lamp can cool themselves only by radiation and convection. Clothing serves to reduce heat loss by conduction, convection, and radiation.

An object heated to incandescence becomes brighter as its temperature increases. The Stefan-Boltzmann law states that the power (P) emitted per unit area is proportional to the fourth power of the temperature of the blackbody (T),

$$P \propto T^4. \tag{2}$$

Notice that this power is not the total power emitted over the entire surface of the blackbody, but is the power from each unit area of the surface.

The **luminosity** (L) of a star is the total power emitted over the entire surface of the star. The power emitted by the star per unit area, given in Equation 2, can then be multiplied by the surface area of the star to get the star's luminosity. Since the area of a sphere is proportional to the square of the radius (R), the luminosity relation becomes

$$L \propto R^2 T^4. \tag{3}$$

Thus, the radius of the star is related to its luminosity and its temperature by

$$R \propto \frac{\sqrt{L}}{T^2}. \tag{4}$$

Example 3: Rigel (Beta Orionis) is about 57,000 times more luminous than the Sun and its temperature is about 2.0 times that of the Sun. How many times larger is Rigel than the Sun?

$$\frac{R_{Rigel}}{R_{Sun}} = \frac{\sqrt{L_{Rigel}/L_{Sun}}}{\left(T_{Rigel}/T_{Sun}\right)^2} = \frac{\sqrt{57000}}{(2.0)^2} = 60.$$

If the Sun were to swell to the size of Rigel, it would reach almost to the orbit of Mercury.

In 1964, Arno Penzias and Robert Wilson, two scientists working at Bell Telephone Laboratories in New Jersey, discovered radio radiation coming from all regions of the sky. When the intensity of this radiation is plotted against wavelength, it approximates the blackbody curve of an object at a temperature of about 3 K. What Penzias and Wilson had discovered are the remains of the **Big Bang**, the explosion that formed the Universe. This radiation is now called the **cosmic background radiation**.

Astronomers have realized since the 1940's that the early Universe must have been very dense and hot, and that it has been expanding and cooling since its formation. This time in the early Universe is called the Big Bang, since it was the explosive beginning from which the history of the Universe is traced. Shortly after the Big Bang, the Universe was filled with **gamma rays**, high energy electromagnetic radiation. As the Universe has continued to expand and cool, the peak in the cosmic background radiation has continued to shift to longer wavelengths. The peak wavelength is currently in the radio portion of the electromagnetic spectrum.

In 1989, the Cosmic Background Explorer satellite was launched by NASA to accurately measure the cosmic background radiation. Results from COBE show that the cosmic background radiation is remarkably uniform in all directions and matches the blackbody curve for a temperature of 2.735 K. The uniformity of the radiation tells astronomers that when the Universe was younger than a million years it was very uniform in temperature and density. More precise measurements have confirmed small-scale fluctuations in the early Universe that led to the formation of galaxies.

Procedures

Apparatus
nothing.

A. Stefan-Boltzmann and Wien's laws

1. If the temperature of an object increases by a factor of 4, how does the peak wavelength change?

2. If a star is the same size as the Sun but its temperature is twice that of the Sun, how many times more power is emitted by the star than the Sun?

3. If a star is the same temperature as the Sun but has twice the diameter, how many times more power is emitted by the star than the Sun?

4. What is the peak wavelength of the cosmic background radiation? Hint: a blackbody at a temperature of 6000. K produces a peak wavelength of 480 nm.

B. Stars on the Main Sequence

1. According to Table 1, a class O5 main sequence star has a peak wavelength of 72.4 nm. Use Wien's law to compute the temperature of this star.

2. According to Table 1, a class O5 main sequence star is 501,000 times more luminous than the Sun. Compute the approximate radius of this star using the Stefan-Boltzmann law.

3. Repeat these computations for each of the spectral classes listed in Table 1.

C. Some Nearby or Bright Stars

1. Table 2 lists the luminosities (relative to the Sun) and peak wavelengths (in nanometers) of several stars. Construct a table of surface temperatures for these stars. Hint: a 6000. K star produces a peak wavelength of 480 nm.

2. Compute the radii of the stars in Table 2 in terms of the Sun's radius. Arrange these results in a column of your table.

3. Convert these radii to astronomical units. Arrange these values in a column of your table.

4. For each of the stars in Table 2, tell whether Earth would be inside the star if it were placed where the Sun is.

Table 1: Some properties of main sequence stars

Spectral Type	Luminosity (Sun = 1)	Peak Wavelength (nm)
O5	501000	72.4
B0	20000	100
B5	790	190
A0	79	290
A5	20	340
F0	6.3	390
F5	2.5	440
G0	1.3	480
G5	0.8	520
K0	0.4	590
K5	0.2	700
M0	0.1	830
M5	0.01	1000
M8	0.001	1200

Table 2: The luminosities and peak wavelengths of several stars

Name	Luminosity (Sun = 1)	Peak Wavelength (nm)
Antares A (α Sco A)	5500	930
Arcturus (α Boo)	115	690
Barnard's Star (in Oph)	0.00045	1200
Betelgeuse (α Ori)	14000	930
Canopus (α Car)	1510	410
Fomalhaut (α PsA)	13.8	310
Pollux (β Gem)	41.7	650
Rigel Kentaurus A (α Cen A)	1.56	480
Rigel A (β Ori A)	46000	240
Spica (α Vir)	2400	110
Vega (α Lyr)	55.0	290
Wolf 359 (in Leo)	0.000018	1200

Binary Stars

Many stars do not travel alone, but have companion stars. A binary star system consists of two stars that orbit each other. Information can be obtained about the stars in a binary system, such as their masses, which is not readily discernible from single stars.

Stars that are not gravitationally bound but appear close to each other from the observer's vantage point are called **optical doubles**. These are chance alignments and provide no more useful information about stellar properties than single stars. Optical doubles seen from Earth would not appear as optical doubles from most other places in the galaxy.

Binary star systems consist of two stars that are gravitationally bound to each other and orbit a common center. These are sometimes called **physical doubles**. If both components of the system can be seen as separate stars, then the system is called a **visual binary**.

The brighter star in a binary system is called the **primary** and the dimmer one is the **secondary**. Most of the known binary star systems were named as single stars before they were discovered to be binary. When a system is found to be binary, the primary is designated with an 'A' and the secondary is designated with a 'B'. So, the binary star system, Mesarthim (Gamma Arietis), contains two stars, Mesarthim A and Mesarthim B. The Bayer Constellation Designation (BCD) of the system is adjusted similarly, as Gamma Arietis A and Gamma Arietis B. An alternate naming system superscripts an Arabic numeral over the Greek letter of the BCD, '1' for the primary and '2' for the secondary, as γ^2 Arietis for the secondary of this system.

The apparent distance between the primary and secondary is called the **separation**, and is typically measured in seconds-of-arc. The orientation of the secondary to the primary is called the **position angle**. The position angle is 0° when the secondary is due north of the primary, 90° when due east, 180° when due south, and 270° when due west. The amount of time it takes the stars to orbit each other is called the system's **period**.

Some care must be taken when determining the orbits of stars about each other, as the orbital planes are likely to be inclined from our point of view on Earth. A circular orbit, for example, will appear eccentric from every angle other than 90° to the plane of the orbit.

In some binary systems, only one star is visible while the companion is hidden in the glare of the primary. As the two stars orbit each other, the primary can be seen to wobble periodically. Some characteristics of the two stars, like their masses, can be determined by carefully measuring the wobble of the primary. Binaries that are discovered in this way are called **astrometric binaries**.

Spectroscopic binaries are those binaries that cannot be resolved visually as two stars but show their dual nature only through their spectra. Edward C. Pickering first noticed this in 1889 when he observed the spectrum of the "star" Mizar (Zeta Ursae Majoris); he saw that the star's spectral lines were doubled. As he continued to observe, the positions of the doubled spectral lines shifted, periodically becoming single. This spectral shifting is caused by the Doppler effect. As the companion star moves in its orbit away from the observer, its spectral lines are Doppler shifted to longer wavelengths, and when it moves towards the observer, its lines are Doppler shifted to shorter wavelengths.

The star Algol (Beta Persei) is an excellent example of another category of binary stars, called **eclipsing binary stars (EBS's)**. Algol changes its brightness by almost 300% about every three days. To may of the ancients, any change in the "constancy" of the heavens was a portent of evil. The name, "Algol," is derived from the Arabic, "Al Ra's al Ghul," which means "The Demon's head." Other civilizations have also identified Algol with similar terms: the Chinese called it "Tseih She," which means "Piled-up corpses" and to the Hebrews it was "Rosh ha Satan" or "Satan's head." Even today, Algol is sometimes called "the demon star" or "the winking eye of the demon."

John Goodricke first determined the period of Algol in 1872, and offered an explanation: the variations in brightness are due to another body orbiting Algol and periodically eclipsing the star. Not all astronomers of the time agreed with this, in fact, Sir William Herschel was one of the biggest doubters because he had observed Algol on many occasions and had never seen another object. Finally in 1889, well after Goodricke's untimely death, astronomers observed two separate spectra for Algol, thus confirming Goodricke's hypothesis.

EBS's can only be observed when a binary star system's orbit is close to edge-on from the observer's point of view. A plot of the system's brightness versus time is called a **light curve**, and is a very useful tool in the study of EBS's. Since there are two stars in the system, there are two reductions or dips in brightness. Such reductions in brightness are called **minima**; the more significant dip is called the **primary minimum** and the less significant dip is the **secondary minimum**. The interval between successive primary minima is the EBS's period.

Figure 35-1 shows the orbit and Figure 35-2 shows the light curve of a hypothetical EBS. In this system, the brighter primary star orbits the more massive, but cooler secondary star. When the primary star is in position 1, it is behind the secondary star and in eclipse; thus the EBS appears faintest as shown by the primary minimum of the light curve. The secondary minimum occurs when the primary star is at position 5; some of the light of the secondary star is eclipsed so a slight drop in brightness occurs. The EBS appears brightest from positions 2 through 4 and again from 6 through 8, where light from both the primary and secondary stars is visible.

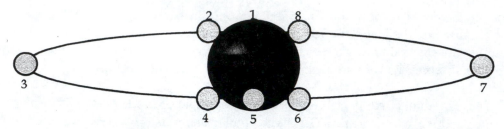

Figure 35-1: An eclipsing binary star system. Some orbital positions are labeled and correspond to labels on the light curve in Figure 2. In this figure, the dimmer, cooler, secondary is fixed in the center, but in reality, both stars orbit the center of mass of the system.

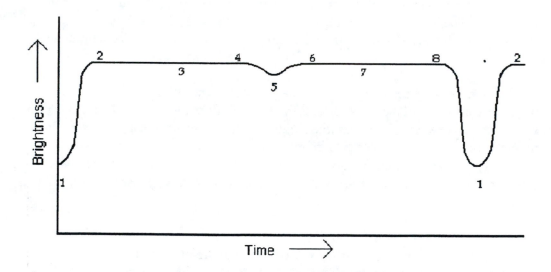

Figure 35-2: The light curve of an eclipsing binary star system. The labels on the curve correspond to position labels in Figure 35-1. Such light curves are repeated with each orbit.

The magnitude of an EBS at a particular time can be determined by visually comparing the apparent brightness of the EBS to nearby stars whose magnitudes are known. For example, one nearby star may be 6.3 magnitude, a second 6.7 and a third 6.9. If the EBS is dimmer than the 6.3 magnitude star and brighter than the 6.7 magnitude star, one could estimate the EBS to be 6.5 or 6.6 magnitude. An observer using a photometer, which measures light intensity, would typically calibrate that instrument on several stars of known magnitude and then record measurements of the EBS.

An EBS is an **extrinsic** variable, since the variation in the system's brightness is due to the external phenomenon of one star eclipsing the other. An **intrinsic** variable, such as a Cepheid, varies in brightness due to changes within the star. See Exercise 36, Variable Stars, for more information.

Procedures

Apparatus

telescopes, binoculars, and star charts or computer planetarium.

The exercises for this lab are divided into two sections, Visual Binaries and EBS's. The sections may be separately and done in any order.

Visual Binaries

A. Pre-Observing

1. Use a detailed star chart or computer program and the list of visual doubles found in Table 1 to determine which visual doubles will be high in the sky during the observing session.

2. Make a sketch of the region around each visual binary, showing stars that you can use as guides to find the binary.

B. Observing

1. Help set up the telescopes and other equipment. Note in your report the activities you perform. Remember to record the dates and times of all observations.

2. Locate the first double star on your list with binoculars or a telescope.

3. Describe the two stars in the system. Include the colors of the stars, the relative brightnesses of the stars, an estimate of the apparent separation of the stars, the position angle, and the ease with which you were able to "split" the stars (that is, see each star individually). Close doubles may require higher magnification to be split.

4. Repeat these observations with the two other double stars on your list.

5. Help put away equipment. Note on your report the activities you perform.

Eclipsing Binaries

These exercises require a minimum of two observing sessions separate by about half the selected EBS's period. More observing sessions would increase the probability of observing during a minimum.

A. Pre-Observing
1. Use a detailed star chart or computer planetarium and the list of EBS's found in Table 2 to determine which EBS's will be high in the sky during the observing session. Choose at least one to observe.

2. The EBS should be observed at least twice in order to observe variations in brightness. Determine when to observe the EBS from its period. For example, if the EBS has a period of 2.5 days, then you should observe it 1 day after your first observation, so that the EBS will be at a different point in its light curve and will appear brighter or dimmer. If it is overcast on one of the scheduled days, another observing session will be needed.

3. Make a sketch of the region around the EBS to help you in locating and identifying the EBS.

4. Record the magnitudes of the stars in your sketch. You will use these magnitudes to estimate the current magnitude of the EBS.

B. Observing
1. Help set up the telescopes and other equipment. Note in your report the activities you performed. Remember to record the dates and times of all observations.

2. Locate the EBS through binoculars or telescope.

3. Estimate the magnitude of the EBS by comparing it with the nearby stars, whose magnitudes are included on your sketch. Record the date and time of your observation.

4. Help put away the equipment, noting in your report the activities you perform.

5. Repeat steps 1, 2, 3, and 4 one or two nights later, depending on the expected period of the EBS.

6. How much time elapsed between your observations?

7. Where in the light curve did you observe the EBS? Was the EBS near its primary minimum during one of your observations?

8. By what percentage did the brightness of the EBS change?

Table 1: Double Stars

Star	Separ.	Magn.	RA/DEC	Comments
γ Ari (Mesarthim)	8"	4.80, 4.80	01:54/+19°18'	An equal white pair some have likened to 'twin diamonds'.
γ And (Almach)	12"	2.30, 5.10	02:04/+42°20'	Slightly orange and greenish blue colors.
α Cvn (Cor Caroli)	17"	2.90, 5.40	12:56/+38°19'	Pale yellow and blue colors.
ζ UMa (Mizar)	15"	2.4, 4.0	13:24/+54°56'	First discovered binary (ca. 1650 by Riccioli). Greenish-white in color. Both stars are also spectroscopic binaries.
δ Apodis	105"	4.80, 5.20	16:20/-78°41'	Red and orange colors.
ε Lyr	206"	5.10, 4.50	18:44/+39°40'	The famous 'Double Double'. Good eyesight can split the principal pair without optical aid. Each double is itself a double (6.02 and 5.37 magnitudes and 2.7" and 2.3" separations).
β Cyg (Albireo)	34"	3.09, 5.11	19:31/+27°58'	Bright, gold and sapphire colors. This may be only an optical double.

Table 2: Eclipsing Binary Stars

EBS	RA	DEC	Magnitude Range	Period
S Ant	09:32	-28° 38'	6.4 - 6.9	0.648 day
δ Aql	19:37	+05°17'	5.0 - 5.2	1.950 days
R Ara	16:40	-57° 00'	6.0 - 6.9	4.425 days
R CMa	07:20	-16° 24'	5.7 - 6.3	1.136 days
RZ Cas	02:49	+69° 38'	6.2 - 7.7	1.195 days
U Cep	01:02	+81° 53'	6.7 - 9.2	2.493 days
δ Lib	15:01	-08° 31'	4.9 - 5.9	2.327 days
β Lyr	18:50	+33° 22'	3.3 - 4.3	12.940 days
U Oph	17:17	+01° 13'	5.9 - 6.6	1.677 days
β Per (Algol)	03:08	+40° 57'	2.1 - 3.4	2.867 days
U Sge	19:19	+19° 37'	6.6 - 9.2	3.381 days
RS Sgr	18:18	-34° 06'	6.0 - 6.9	2.416 days
HU Tau	04:38	+20° 41'	5.9 - 6.7	2.056 days

Variable Stars

Stars that vary noticeably in brightness are called **variable stars**. Stars that vary in brightness due to some internal cause are referred to as **intrinsic variables**. **Eclipsing binary stars (EBS's)** vary in brightness because one star is eclipsed by a second orbiting star. EBS's are said to be **extrinsic**, since their brightness changes are not due to an internal event. This lab considers three main classes of variable stars: pulsating, eruptive, and eclipsing.

The range of change in brightness is called the variable star's **amplitude**. The amplitude depends on the type of variable star and can range from as little as a fraction of a magnitude up to several magnitudes. A plot of the change in magnitude over time is called a **light curve**. The peaks in the light curve correspond to peaks in brightness and are called **maxima**. The low points in the light curve are called **minima**.

Pulsating variables are stars that periodically expand and contract. The major types of pulsating variables are cepheid, W Virginis, RR Lyrae, dwarf cepheid, Beta Canis Majoris, long-period, semi-regular, irregular, RV Tauri.

Cepheid variables (named after the prototype, Delta Cephei) are stars that pulse with a clock-like regularity. Cepheid periods range from about one day to several weeks. Cepheids are important in that their periods of variation are directly proportional to their absolute luminosities. This relationship allows astronomers to know the intrinsic brightness (or absolute magnitude) of an observed cepheid no matter where it is. The distance to the star can be determined by comparing its absolute magnitude with its apparent magnitude. This technique has proved invaluable in determining the distances to galaxies whose cepheids can be observed. Classical cepheids are young, metal-rich, Population I stars that mostly populate the arms of spiral galaxies. Classical cepheids are typically F- and G-type stars.

The expansion and contraction of cepheids are caused by the build-up and release of internal pressure. Deep within the star lies a layer of ionized helium, compressed by the outer layers. When the ionized helium is compressed, it becomes opaque, which causes it to absorb more radiation from the core. This raises the layer's temperature, which causes it to expand outward. The pressure is released when the star expands, thus allowing the star to contract and the process to start again. The location of the ionized helium layer is crucial, for if the layer is too near the core or too near the photosphere, the star will not pulsate. Most pulsating variables are also driven by ionization layers.

W Virginis variables were originally classified as cepheids. But, W Virginis variables are older and dimmer than classical cepheids, and have a slightly different period-luminosity relationship. W Virginis variables are metal-poor, Population II stars, and are typically found in galactic halos. Like classical cepheids, W Virginis variables are mostly of the F and G spectral classes.

RR Lyrae variables are smaller and dimmer than cepheids, have periods of a day or less, and have amplitudes of about half a magnitude. RR Lyrae variables all have absolute magnitudes of about +0.5; once a star is determined to be an RR Lyrae variable by its light curve, its absolute magnitude is known and may be used as an indicator of distance. RR Lyrae variables are Population II, giant stars of the A and F spectral classes.

Dwarf cepheid variables resemble RR Lyrae stars but with smaller amplitudes (only a few tenths of a magnitude) and shorter periods (only a few hours). They are mostly subgiant and bright main sequence stars of the A and F spectral classes. SX Phoenicis is a well known dwarf cepheid.

Beta Canis Majoris, sometimes called *quasi-cepheids*, are bright, giant stars of the spectral classes B1, B2, and B3 that exhibit a cepheid-like period-luminosity relationship. Their periods range from about 3.5 hours to about 6 hours. Beta Canis Majoris variables have only slight amplitudes, up to about 0.2 magnitude. Several of these variables are found in groups of young stars, and may indicate that quasi-cepheids are young, massive stars just evolving off the main sequence. This idea is supported by the fact that they exhibit a gradual lengthening of the period of pulsation, which is expected with the slow expansion of stars just leaving the main sequence.

Long-period variables are sometimes called *Mira variables*, after the prototype star, Omicron Ceti. They are red giants with periods between about 100 and 700 days. The periods and amplitudes are somewhat variable, caused by shock waves originating deep within the star and are quite dramatic. The long-period variable, Mira (Omicron Ceti), goes from about third magnitude to ninth magnitude in eleven months. Long-period variables are found in both Population I and Population II and are of the M, R, N, and S spectral classes.

Semi-regular variables are red giant stars, typically of the M spectral class. They have poorly defined periods, often with unpredictable magnitude variations. Alpha Herculis and W Cygni are archetypes of this class of pulsating variable.

Irregular variables are giant stars of various spectral classes that exhibit unpredictable changes in brightness. Betelgeuse (Alpha Orionis) and Antares (Alpha Scorpii) fall into this type of pulsating variable.

RV Tauri variables are bright giant stars whose sequence of variations consists of a shorter period of 30 to 150 days superimposed on a longer period of 3 or 4 years. RV Tauri variables are characterized by having two equal maxima and two unequal minima during its short-period variation. Amplitudes may be up to about three magnitudes. RV Tauri variables are typically G- or K-type stars.

Eruptive variables consist of stars that brighten or dim suddenly and, usually, unpredictably. The major types of eruptive variables are novae, dwarf novae, recurrent novae, supernovae, flare stars, R Coronae Borealis variables, and T Tauri variables.

Novae are stars that suddenly and explosively brighten. Fast novae brighten by a dozen or more magnitudes in only a day or two. They may reach absolute magnitudes of -8 to -10. They then drop by about three magnitudes in a month, but may take years to return to their pre-explosion magnitudes. Slow novae also brighten quickly, but only to magnitudes of about -6. It may then take them six months to drop three magnitudes, and, perhaps, decades to return to their pre-explosion magnitudes.

Novae occur in binary star systems where one star is more evolved than the other, typically a white dwarf. The two stars are often so close to each other that material (mostly hydrogen) pours off the younger star into an **accretion disk** around the white dwarf. This material slowly descends to the surface where it becomes highly compressed and heated by the white dwarf's gravity. Eventually, the layers of hydrogen become hot enough that runaway thermonuclear explosions blow the built-up layers into space with speeds far exceeding the escape velocities of the system. Gaseous clouds, **nova remnants**, can be seen expanding away from the stars. The white dwarf interiors are very dense and removed from the thermonuclear explosions, so they remain intact after the explosion. Thus, the nova may repeat when enough hydrogen has been built-up on the white dwarf's surface, though this may take hundreds of thousands of years.

Dwarf novae (or *SS Cygni variables*) exhibit unpredictable, rapid brightenings several times a year. Within a day or two, the dwarf nova system may brighten by four or five magnitudes. The brightenings seem to be caused by explosions in the accretion disk rather than on the surface of the more evolved star.

Recurrent novae are those that have shown two or more outbursts. They have smaller amplitudes than standard novae, and they return to normal brightness more quickly. They also exhibit irregular variations in brightness between outbursts. T Coronae Borealis is typical of this class of variable. Its normal magnitude is about +10, but, in 1866, it rose in magnitude to +2. The star dropped to its former magnitude in about 8 days after maximum. About three months later, the star increased a couple of magnitudes in brightness, but subsided after about three more months. In 1946, T Coronae Borealis gave a repeat performance, proving that it is a recurrent nova.

Supernovae are the explosive deaths of stars. Supernovae show sudden rises in brightness (usually less than a day), and may increase in brightness by 20 or more magnitudes. Supernovae come in two types, labeled Type I and Type II. Type I supernovae involve a binary star system in which a companion dumps material onto a white dwarf. At some point, enough mass has been dumped onto the star's surface to cause carbon to begin to fuse in its core. The carbon fusing moves outward from the core at enormous speeds. If those speeds exceed the speed of sound, the star blows apart. Type II supernovae involve the death of a single, massive star as it runs out of fuel in its core. See Exercise 33, Elements and Supernovae.

Flare stars are faint red dwarfs that typically increase in brightness by one or two magnitudes in only a minute, and then fade back to normal within a few minutes. UV Ceti and Alpha Centauri C are typical of this class of variable. These flares are caused by spectacular stellar flares (similar to solar flares). They are noticeable mainly because the stars are so dim to begin with.

R Coronae Borealis (or *R CrB*) *variables* might be thought of as "reverse-novae." They are characterized by extremely long maxima followed by sudden and unpredictable drops in magnitude. The prototype for this class of variables, R Coronae Borealis, normally shines at fifth magnitude, but will suddenly drop to about thirteenth magnitude. Over the next few months, R Coronae Borealis will slowly and irregularly climb its way back to fifth magnitude. These variations are caused by the sudden condensation of carbon dust grains, that are slowly dissipated by the stars stellar wind.

T Tauri variables are young protostars with masses between 0.2 and 2.0 solar masses. They have not yet reached the main sequence stage and are usually surrounded by dense nebulosity. T Tauri variables may brighten greatly for a few years, then remain at that brightness for a while before plunging to dimmer magnitudes at some unpredictable time. The reasons for T Tauri variability are not fully understood, and may involve infalling matter from the surrounding nebulosity, stellar flares caused by strong magnetic fields, changes in the rate of nuclear fusion as the protostar settles towards the main sequence, or simply the surrounding gas being blown away and momentarily revealing the true brightness of the star.

Eclipsing variables (or **Eclipsing Binary Stars** or **EBS's**) are found in binary systems in which one star periodically passes in front of another as seen by a remote observer. If each star of the system has a different magnitude, then the brighter star is called the **primary** and the dimmer star is called the **secondary**. More information and some exercises concerning EBS's can be found in Exercise 35, Binary Stars. There are several classes of eclipsing variables: Algol systems, Lyrid systems, dwarf eclipsing systems, ellipsoidal variables.

Algol systems are widely-spaced binary systems in which one star is seen to move periodically in front of another. Because the stars are widely spaced, the light curve remains fairly flat between eclipses. Often, one star of the system is dimmer than the other. In this case, the light curve exhibits asymmetrical dips as first the brighter component and then the dimmer component are eclipsed.

The archetype of this class of EBS is Algol (Beta Persei). One of the peculiarities of the Beta Persei system, which is shared by many other EBS's of this class, is that the dimmer, less massive secondary star has evolved off the main sequence while the brighter, more massive primary remains on the main sequence. This is a paradox, for if the stars began at the same age, which would be expected as they condensed from the same nebula, the more massive star would evolve faster. It seems now that the less massive secondary was originally the more massive star. As it evolved, the originally more massive component expanded and dumped large quantities of matter onto its less massive companion, making it now the more massive component. Similar phenomena may occur in Lyrid systems as well.

Lyrid systems are very close, usually giant stars that revolve about each other. Each component of a Lyrid system is distorted into an ellipsoid due to the strong gravitational pull of its companion and centrifugal action of its rapid revolution. The light curve of a Lyrid system is sinusoidal with few flat areas. The archetype of this class is the Beta Lyrae system. In this system, the two stars are connected by an enormous filament of plasma flowing from the larger star to the smaller one.

Dwarf eclipsing systems (or *W Ursae Majoris stars*) consist of rapidly revolving stars that are nearly in contact with each others. These stars are so close that periods are often less than a day.

Ellipsoidal variables are binary systems that do not actually eclipse each other, but vary in light intensity due to the changing amount of luminous surface seen from Earth as the stars revolve. Zeta Andromedae is a typical ellipsoidal variable.

Finally, there are some variables that can not be conveniently placed in any of the major classes. As similarities are found among these variables, new classes may be developed. One possible new group would be **nebular variables**, whose members vary due to some type of interaction with surrounding nebulosity.

For most variable stars, observations are tedious and time consuming. Most variables are not bright enough to see with the unaided eye throughout their periods, so binoculars or telescopes are needed.

A key observational activity involves estimating the brightness of the variable. Professional astronomers use electronic photometers to measure accurately and precisely the amount of light received from the variable. However, there is a simple method for obtaining a rough estimate of the variable's magnitude visually. A detailed star chart or computer planetarium program can be used to determine the magnitudes of stars appearing near the variable. The observer needs simply to compare the brightness of the variable with that of the nearby stars to get an estimate of the variable's current brightness. For example, one star near the variable may be at magnitude 5.1, a second may be at magnitude 5.5, a third may be at magnitude 5.7, and so on. If the brightness of the variable is halfway between the first and second stars, then it can be estimated to be at magnitude 5.3.

The accuracy of the visual estimate depends on the range of stars nearby with which the variable star is compared, the observing conditions (haze, light pollution, and the Moon can interfere the observer's perception of the variable's brightness), and the color of surrounding stars (a reddish 6.0 magnitude star will appear different to the eye than a bluish 6.0 magnitude star).

Procedures

Apparatus
Pre-observing: star charts or computer planetarium program.
Observing: various telescopes and eyepieces, binoculars.

A. Pre-Observing
1. Look over a list of variable stars, such as Table 1. Use a computer planetarium program, almanac, or star wheel to determine which are visible and high in the sky around the expected dates and times of the observing sessions. It will be especially helpful if several different variables can be found that are close enough to each other in the sky that they may be observed during one observing session.

2. Determine the number of observing sessions that will be used. Variables with periods of one or two days can be observed during a week, witnessing perhaps several maxima and minima. Variables with periods measured in weeks may be used for term projects.

3. After selecting three variables to observe, use a detailed star chart or computer planetarium program to locate neighboring stars and their magnitudes. Make sketches of the regions around the variables, labeling the magnitudes of the nearby stars. If possible, also label the expected magnitude ranges of the variables.

4. Determine what equipment will be needed at the observing site, such as binoculars, telescopes, eyepieces, etc. Make sure that you understand the use and operation of each item.

B. Observing
1. Help set up the equipment at the observing site. Briefly note in your report the activities you performed.

2. Locate the target variable star through binoculars or telescope.

3. Compare the brightness of the variable star with the neighboring stars. Estimate the current magnitude of the variable. Record this value along with the date and time.

4. Repeat steps B.2 and B.3 for each selected variable.

5. Repeat steps B.1 through B.4 for each observing session.

6. Help put away the equipment. Briefly note in your report the activities you performed.

C. Post-Observing

1. Although it is not possible to develop complete light curves for the observed variables, it might prove interesting to represent the collected data graphically. Make a plot of magnitude (on the y-axis) versus time (on the x-axis) for each variable.

2. For each variable, list the maximum and minimum magnitudes that you observed.

3. Based on the published information about the selected variables, did you observe any of the variables near a maximum or minimum?

4. Are your observations consistent with the published information about the selected variables? Explain.

Table 1: Selected Variable Stars

Variable	Type	RA/DEC (epoch 2000.0)	Magnitude Range	Period
δ Cephei	cepheid	22:29/+58°23'	3.90 - 5.09	5.37 days
η Aquilae	cepheid	19:52/+00°59'	4.08 - 5.36	7.18 days
ζ Geminorum	cepheid	07:04/+20°35'	3.68 - 4.16	10.15 days
α Ursae Minoris (Polaris)	cepheid	02:24/+89°14'	1.94 - 2.05	3.97 days
X Sagittarii	cepheid	17:47/-29°50'	4.24 - 4.80	7.01 days
β Doradus	cepheid	05:34/-62°30'	4.03 - 5.07	9.84 days
W Virginis	W Vir	13:24/-03°07'	9.5 - 10.7	17.3 days
SW Andromadae	RR Lyrae	00:21/+29°07'	9.3 - 10.3	0.4 day
SX Phoenicis	dwrf cepheid	23:46/-41°36'	7.1 - 7.5	79 minutes
β Canis Majoris	β CMa	06:23/-17°57'	1.99 - 2.02	6 hours
O Ceti (Mira)	long-period	02:19/-03°01'	3.4 - 9.0	332 days
α Herculis	semi-regular	17:14/+14°24'	3.1 - 3.9	90 days
μ Cephei	irregular	21:42/+58°33'	3.6 - 5.1	irregular
R Scuti	RV Tauri	18:45/-05°46'	4.9 - 8.2	144 days
SS Cygni	dwrf nova	21:41/+43°30'	8.1 - 12.1	20 - 90 days
R Coronae Borealis	R CrB	15:47/+28°19'	5.8 - 14.4	irregular
β Persei (Algol)	Algol EBS	03:08/+40°56'	2.1 - 3.4	2.8 days
β Lyrae	lyrid	18:50/+33°22'	3.4 - 4.1	13 days

Astronomy Laboratory Exercise 37
Astronomy Math Review

The physical sciences deal with the measurement of phenomena. Scientists must be able to handle numbers comfortably and understand the significance of their measurements. This lab reviews some of the basic mathematics used in this text.

A. Scientific Notation or Powers of Ten

Any number can be written as a real number between 1 and 10 times ten raised to an integral power. That is, the number can be written in the form $X \cdot 10^n$, where X ranges from 1 to 10 and n ranges from -infinity to +infinity. Note that X can not be 10 and n can not be infinite.

Example 1: Write four and a half billion years and 18 billionths of a second in scientific notation.

A billion is 1,000,000,000, that is a one followed by nine zeros. The exponent is the number of times the decimal point must be shifted to the left to be placed to the right of the one. So, a billion can be written as 10^9. Thus, four and a half billion years is $4.5 \cdot 10^9$ years.

A billionth is 0.000000001, that is a one preceded by eight zeros before the decimal point. The exponent can be found by simply counting the number of times the decimal point must be shifted to place it on the right side of the one. Shifts to the right produce negative exponents. So, a billionth is 10^{-9} and eighteen times that is $18 \cdot 10^{-9}$. But this is not in proper form, so the decimal point is shifted to the left and the exponent is increased by the number of shifts to the left. In this case, 18 billionths of a second is $1.8 \cdot 10^{-8}$.

B. Significant Figures

Significant figures are the known or certain digits in a measurement. The number of significant figures indicates the precision of the measurement. More precise measurements have more significant figures. The precision of measurements is determined by the instruments used to take them.

Example 2: You use a centimeter ruler (a ruler with marks for centimeters only) to measure the length of an object, and determine it to be 10.3 centimeters long. This number has three significant figures with the rightmost figure (the three millimeters) being an estimate. When the measurement is redone with a millimeter ruler (a ruler with marks for millimeters and centimeters), the length is 10.34 centimeters. The object did not change length, but the precision of the measurement did. The measurement with the millimeter ruler is more precise because you were able to confirm the 3 millimeters and estimate the 4 tenths of a millimeter.

The following rules are used in writing significant figures:
1. all *non-zero integers* are significant
2. *leading zeros* (those that precede all other digits) are NOT significant
3. *captive zeros* (those that fall between other significant digits) are significant
4. *trailing zeros* (those that follow all other digits) are significant only if they follow the decimal point
5. a zero followed by only an *explicitly written decimal point* is significant.
6. In *addition* and *subtraction*, retain as significant only digits to the left of the first non-significant digit in the least precise quantity that went into the calculation.
7. In *multiplication* and *division*, retain only as many significant figures in the result as appeared in the least precise quantity.

Note: calculations should be performed without regard to significant figures, then the result should be rounded to the appropriate number of significant figures.

Example 3: How many significant figures are in 0001200300.040005000?
All leading zeros are not significant, so the first three zeros do not count. All captive zeros count. Trailing zeros count only if they follow the decimal point, which they do here. So, this number can be rewritten without loss as 1200300.040005000. Thus, there are 16 significant figures in this number.

Example 4: How many significant figures are in 12000. and 12000?
12000. has two non-zero figures and three trailing zeros. These zeros are significant since they are followed by an explicitly written decimal point. Thus, there are five significant figures in this number.
12000 has only two significant figures.

Example 5: Here are some examples using arithmetic operations.
4.56 (three sig.fig.) * 1.4 (two sig.fig.) = 6.38 ==> 6.4 (two sig. fig.)
12.11 + 18.0 (first decimal place) + 1.013 = 31.123 ==> 31.1 (first dec. place)
1.45 * (21 + 20) = 59.45 ==> 60 (20 has one sig. fig.)
1.45 * (21 + 20.) = 59.45 ==> 59 (20. has two sig. fig.)

Note: numbers written in scientific notation must have the correct number of significant figures. Thus, $1.20 \cdot 10^3$ has three significant figures, while $1.2 \cdot 10^3$ has only two.

C. Percentage Error and Percentage Difference

Percentage error is a measure of the closeness of a calculated or observed (Obs) value to the published or actual (Act) value. In this way, percentage error is a quantification of accuracy and is given by the following formula, where the vertical lines indicate absolute value:

$$\%err = \left| \frac{Obs - Act}{Act} \right| *100\% . \tag{1}$$

Notice that the absolute value lines cause the percentage error always to be positive.

Percentage difference is a measure of how close two measurements (meas1 and meas2) are to each other, or the precision of the measurements. It is given by:

$$\%diff = \left| \frac{meas1 - meas2}{\left(meas1 + meas2 \right)/2} \right| *100\% . \tag{2}$$

As with percentage error, percentage difference is always positive.

Example 6: The mass of an object is measured to be 1.78 grams, but the manufacturer published 1.80 grams as the mass. What is the percentage error between our measurement and that of the manufacturer?

Since we are comparing our value to that of the manufacturer, we shall take 1.80 grams as Act. Thus,

$$\left| \frac{1.78 \cdot g - 1.80 \cdot g}{1.80 \cdot g} \right| *100\% = 1.11\% .$$

Example 7: The mass of an object is measured twice to be 1.78 grams and 1.81 grams. What is the percentage difference between these two measurements?

It does not matter which we choose to be meas1 and meas2. Thus,

$$\left| \frac{1.78 \cdot g - 1.81 \cdot g}{\left(1.78 \cdot g + 1.81 \cdot g \right)/2} \right| *100\% = 1.67\%$$

D. Arithmetic Mean

The best way to make good measurements is to make many measurements and calculate the arithmetic mean of those measurements. The arithmetic mean is given by:

$$mean = \frac{meas1 + meas2 + \cdots + measN}{N} , \tag{3}$$

where meas1, meas2, etc. are the measurements and N is the number of measurements.

Example 7: A table is measured to be 1.81 m, 1.83 m, 1.792 m, 1.78 m, and 1.81 m long. The arithmetic mean of the length is 1.8044 m $\Longrightarrow$ 1.80 m (three sig. fig.).

E. Trigonometry and Geometry

A good understanding of the basics of trigonometry is a necessity in astronomy.

Angles will be measured in degrees, minutes-of-arc, seconds-of-arc, and in radians.
$$1° = 60' \text{ (minutes-of-arc)} = 3600'' \text{ (seconds-of-arc)}.$$
$$180° = \pi \text{ radians (rad) or 1 radian is approximately 57.3 degrees.}$$

Much of trigonometry is based on the right triangle.

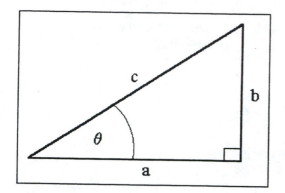

$$c^2 = a^2 + b^2 \qquad : \text{Pythagorean Theorem}$$

$$\sin\theta = b/c \qquad : \text{sine}$$
$$\cos\theta = a/c \qquad : \text{cosine}$$
$$\tan\theta = b/a \qquad : \text{tangent}$$

$$\theta = \arcsin(b/c) \qquad : \text{arcsine}$$
$$\theta = \arccos(a/c) \qquad : \text{arccosine}$$
$$\theta = \arctan(b/a) \qquad : \text{arctangent}$$

When the angle is very small and measured in radians, both $\sin\theta$ and $\tan\theta$ can be approximated by θ. This is called the **small angle approximation**.

Here are some geometrical formulae. **r** is the radius of the circle or sphere.
Circle   **Sphere**
arc length = $r\theta$ surface area = $4\pi r^2$
circumference = $2\pi r$ volume = $(4/3)\pi r^3$
area = πr^2

Example 8: On a particular day, you measure the Moon's angular diameter to be 0.51°. If the diameter of the Moon is 3476 km, how far away is it?

Assume that your line of sight to one side of the Moon forms the adjacent side, **a**, of the triangle in the above figure, and that the diameter of the Moon's disk forms the opposite side, **b**. Then the angle subtended by the Moon (its angular diameter) is θ. Rearranging the definition of the tangent function, gives
$$a = b/\tan(\theta) = 3476 \cdot km/\tan(0.51°) = 390500 \cdot km.$$
Thus, the Moon is 390,500 km from your eyes.

Example 9: Phobos, one of Mars' moons, has dimensions of 14 x 11 x 9 km and a semi-major axis (mean distance from the center of Mars) of 9378 km. Mars' radius is 3393 km. What is the largest angle subtended by Phobos at its mean distance as seen from the surface of Mars?

Form a right triangle as was done in Example 8. The distance from the surface of Mars to Phobos, a, is 9378 km - 3393 km = 5985 km. Thus,
$$\theta = \arctan(b/a) = \arctan(14 \cdot km/5985 \cdot km) = 0.13° = 8.0'.$$

Procedures

Apparatus
none.

1. State the number of significant figures in each of the following.
   a) 28.50 _____
   b) 42000 _____
   c) 0.009090 _____
   d) $1.01 \cdot 10^3$ _____

2. Express the following quantities in scientific notation (use proper significant figures).
   a) 300000 _____
   b) 149600000 _____
   c) 0.0020300 _____
   d) 0.938404 _____

3. Perform the following calculations to the correct number of significant figures.
   a) 60.00 * 60 = _____
   b) 4.210 + 2.104 = _____
   c) 2500 / 5.0000 = _____
   d) 63.30 - 21.3 = _____

4. Perform the following calculations to the correct number of significant figures. Express your answers in proper scientific notation.
   a) $(20. \cdot 10^3) * (4.00 \cdot 10^4) =$ _____
   b) $(4.0 \cdot 10^6) * (1.234 \cdot 10^{-6}) =$ _____
   c) $(200 \cdot 10^3) * (3 \cdot 10^{-1}) =$ _____
   d) 0.004 + 0.0000005 = _____

5. Calculate the number of seconds since you were born up to your next birthday to four significant figures. Express your answer in scientific notation. _____

6. A lab group makes the following measurements of a distance:
   Sue: 240.7, 244.0, 237.2, 239.5, 242.4 mm
   Joe: 250.2, 229.3, 242.2, 241.2, 240.5 mm
   Ann: 240.1, 239.8, 237.1, 228.8, 244.4 mm.

   a) Find the arithmetic mean of each person's measurements.
   Sue: _____ mm   Joe: _____ mm   Ann: _____ mm

   b) If the published value is 240.4 mm, find the percentage error for each average found in 6(a).
   Sue: _____ %   Joe: _____ %   Ann: _____ %

c) According to the percentage errors computed in 6(b), who has the most accurate measurements? _____

d) Compute the arithmetic mean of the averages of 6(a). _____

e) What is the percentage difference between Ann's average value and the average computed in 6(d)? _____

7. An angle of $\pi/4$ radians separates two objects in the sky. What is this angle in degrees? _____

8. If a skyscraper subtends an angle of 70.° as seen from the ground 100. meters away, how high is the sky scraper? _____

9. On a particular day, the Moon subtends about 32'24" in Earth's sky. If the Moon is 3476 km across, how far away is it? You may round to four significant figures. _____

10. Europa has a diameter of 3130 km and a mean distance from the center of Jupiter (its semi-major axis) of 671,400 km. What is the mean angular diameter of Europa as seen from the cloud tops of Jupiter? Jupiter has a diameter of 142,800 km. Hint: the moon's distance from the cloud tops of Jupiter is 671,400 km - (142,800 km / 2). _____

11. You observe Uranus to have an angular diameter of $1.8 \cdot 10^{-5}$ radians when its distance from Earth is $2.86 \cdot 10^{9}$ km. Use the small angle approximation to simplify the computation of the diameter of Uranus in kilometers. _____

12. Determine the distance around Earth's equator. Earth's equatorial diameter is 12,756 km. _____

Computer Planetaria

A **planetarium** usually consists of a large room with a domed ceiling onto which may be projected the images of stars, planets, and other celestial bodies. Planetaria are great tools for learning about the heavens, as they can simulate the sky for a variety of locations and times. A **computer planetarium** is a computer program that allows users, from the casual observer to the dedicated professional, to explore the heavens from the desktop. This lab explores some of the typical features and uses of computer planetaria.

As with a traditional planetarium, a computer planetarium is very helpful in learning the constellations and other imaginary patterns in the sky. Besides stars, a computer planetarium may also show the positions of planets, comets, and asteroids. Before such a program can display these things, it must know the observer's location, the date, and the time.

Observing locations on Earth are input by their latitude and longitude coordinates. Most computer planetaria also include databases of cities and their coordinates, and users may select their locations from these databases. Some computer planetaria even allow users to select locations on other planets in the Solar System from which to observe.

The computer planetarium may limit the range of dates for which it will simulate the sky. This is often due to the accuracy with which it computes the positions of celestial objects. If the computer planetarium does not take into consideration the precession of Earth's axis, for example, then the accuracy of the program becomes steadily worse for dates further into the past or future from some reference date.

Time is often entered into the computer as **local zone time**, that is, the time to which all clocks are set in the local time zone. Earth's surface is divided into 24 time zones. The time at the central longitude of each time zone is used as a reference for all the clocks in that time zone. The Royal Astronomical Observatory is located in Greenwich, UK, through which runs the Prime Meridian (longitude 0°). Clocks at the Royal Astronomical Observatory are set to **Greenwich Mean Time (GMT)**, which is also called **Universal Time (UT)** since it is used around the world as a time reference.

It is often useful to be able to convert Universal Time to the local zone time. Since there are 24 time zones, spaced roughly equally around Earth (360°), each time zone reference longitude must be 15° from the next. Earth rotates approximately 15° each hour, so each time zone is about one hour in difference from the next. Also, zone times decrease from east to west around Earth. Table 1 lists some time zones, their reference longitudes, and the differences in hours from Universal Time.

Table 1: Some Time Zones

Designation	Reference Longitude	Difference in Hours from UT*
Atlantic	60° W	-4
Eastern	75° W	-5
Central	90° W	-6
Mountain	105° W	-7
Pacific	120° W	-8
Hawaii-Alaskan	150° W	-10

* '-' indicates earlier than UT.

Example 1: An astronomical almanac predicts that a variable star will reach maximum brightness at 0314 UT tomorrow morning. What time will a clock in the Central Standard Time (CST) zone read at this instant?

CST's reference longitude is 90° W, so
90° / 15° / hour = 6 hours.
Since CST is west of the Prime Meridian, the clock reads 6 hours earlier or 2114 CST the previous night.

Computer planetaria must overcome two major display problems: how to represent stars over a wide range of brightness and how to represent the curving dome of the sky on a flat display screen. The first problem is usually solved the same way as in printing: by changing the sizes of the stellar images. That is, brighter stars are shown larger than dimmer stars. This method is inaccurate, as all stars but the Sun appear as pin-points to the unaided eye, but it allows stellar brightnesses to be represented.

The second problem concerns showing a three-dimensional view in two-dimensions, which can not be done without some distortion. This distortion usually takes the form of inaccurate placement of the stars relative to each other. There are several popular and useful display formats, called **projections**. Some of these are: spherical, polar, and mercator.

A **spherical projection** shows the entire sky as a globe. Stars near the center of the near side of the globe show the least positional distortion, while those near the limbs of the globe show the greatest distortion. Some spherical projections include a line to indicate the location of the observer's horizon. A variation on this type of projection is a hemispherical projection in which only those stars visible to the observer are shown.

A **polar projection** shows the sky as seen from one of Earth's poles. A north polar projection shows the north celestial pole in the center of the display and the celestial equator along the circumference of the display. As with the spherical projection, stars near the center of the display are closest to their true positions, while those near the circumference show the greatest positional inaccuracy.

The **mercator** or **flat projection** shows part of the sky as a flat map. Distortions to stellar positions are minimized by spreading the error throughout the display (not just at the edges) and by showing only a small portion of the sky.

Many computer planetaria provide a **horizon view**, in which the simulated sky shows those stars seen by an observer facing a particular direction. Such a view may utilize a mercator projection or, more often, a hemispherical projection. The program usually allows the user to select the direction of the view. This allows the user to selectively view those objects rising in the east, setting in the west, or transiting the sky in the north or south.

Along with the ability to label the objects in the display, many computer planetaria also include more information about objects in the heavens, such as current coordinates, magnitudes, distances from Earth, and so on. Typically, the user simply selects an object with a pointer and the information is displayed in a window on the screen. A search feature is also typically included, which allows the user to instruct the computer to search for a particular object and report back information about it.

The computer planetarium may also provide an option to show different **coordinate grids** overlaying the sky. The two most common coordinate systems available in this feature are altitude/azimuth and right ascension/declination (see Exercise 2, Coordinate Systems).

Another very useful feature found in many computer planetaria is an **animation** or **time skip** feature. With this feature, the user can have the computer update the display over time to illustrate the motions in the heavens. For example, the user can view a historical, celestial event or track the motions of planets, comets, or asteroids in the sky.

The computer planetarium may include several subsidiary features, such as eclipse and conjunction prediction routines, satellite trackers, and routines for displaying the current positions of the moons of other planets. Finally, the program may include a print feature that allows the user to make a hardcopy (usually on paper) of the display, which may be taken to the observing site and used as a finder chart.

Procedures

Apparatus

computer planetarium software, computer capable of running the software, printer, and specific instructions on their use, latitude and longitude of the Astronomy Lab or designated observing location.

Note: The student should be given specific instructions on the use of the computer planetarium. These may include on-line documentation and help.

A. Finding Current Conditions

1. Set the observer's location to the latitude and longitude of the Astronomy Lab or a designated observing location. Record this location and its coordinates in your lab report.

2. Set the program to the current date and time. Record these values in your lab report.

3. Set the display mode to show those objects currently above the horizon. It may be necessary to change the direction of the display to see the whole sky.

4. Is the Sun currently above the horizon? If so, what are its altitude and azimuth? Also, in which constellation does the Sun currently reside?

5. Is the Moon currently above the horizon? If so, what are its altitude and azimuth? Also, in which constellation does the Moon currently reside?

6. List the planets currently above the horizon along with their altitude and azimuth coordinates and the constellations in which they reside.

7. Select a view to print. Use the computer's labeling feature to label the objects in the display. Use the program's print feature to produce a hardcopy of the display. If a labeling feature is not available, label the objects (Sun, Moon, and planets) represented on the hardcopy by hand.

B. Displaying Projections

1. If you have not done so already, set the observing location, the date, and the time as instructed.

2. Display those stars currently above the horizon. Turn on the constellation line drawing feature, if available. Rotate the display, so as to view sequentially those objects above the eastern, southern, western, and northern horizons. In your report, list those Solar System objects (Sun, Moon, and planets) visible in each view. If available, have the program label the planets in the sky, then print a hardcopy of one display. If an automatic labeling feature is not available, label the hardcopy by hand.

3. Set the program to display a north polar projection. Turn on the right ascension/declination grid lines option, if available. Notice that Polaris (Alpha Ursae Minoris) is very close to the center of the display at +90° declination. Print a hardcopy of this display, and label those Solar System objects (Sun, Moon, and planets) north of the celestial equator at this time.

4. Set the display to a south polar projection. Turn on the right ascension/declination grid lines option, if available. Notice that there is no bright star near the center of the display at -90° declination. Print a hardcopy of this display, and label those Solar System objects (Sun, Moon, and planets) south of the celestial equator at this time. List any Solar System objects on the celestial equator (those that can be seen from both poles).

5. Select a mercator projection of the area of the sky containing the constellation of Orion. Turn on the star labeling option, if available. Each star may have several names and the program may allow you to select which name to use. Follow the instructor's directions in this matter. Produce a hardcopy of the view.

C. Using Animation

1. View Mars' retrograde motion in Earth's sky.
   a. Set the location to the designated observing site. Set the date to 01 May 2003 and the time to midnight.
   b. Set the projection to mercator, turn on right ascension/declination grid lines, turn off all Solar System objects but Mars, turn off all stars dimmer than magnitude +4, and turn off deep sky objects.
   c. Set the following animation (or time skip) options: set the time increment to 5 days, the stop time to 01 December 2003, and have the program mark the track of Mars.
   d. Align the view so Mars appears to the left of the center of the display.
   e. Start the animation. Notice how Mars loops backward then forward again. Record the dates on which Mars switches direction.

2. View the total solar eclipse of 11 July 1991 as seen from Mexico City, Mexico. Record the following times:
   a. when the Moon first begins to cover the Sun (first contact),
   b. when the Moon just covers the Sun (second contact),
   c. when the Moon first begins to uncover the Sun (third contact), and
   d. when the Moon completely uncovers the Sun (fourth contact).
Note that these times are local time for Mexico City.

3. Obtain from the instructor another celestial event to observe. Record any information requested by the instructor.

D. Planning an Observing Session

1. Acquire from the instructor the location, date, starting time, and ending time for your observing session. Record these values in your lab report. Set the computer planetarium to the location and date of the observing session.

2. Set the program to the beginning time of the observing session. Set the program to show the objects above the western horizon at the beginning of the observing session. What Solar System objects will set before the end of the observing session? Earth rotates at roughly 15° each hour. Altitude/azimuth coordinate grid lines may prove helpful in estimating set times.

3. Rotate the view to the eastern horizon, but keep the time to the beginning of the observing session. List what Solar System objects are visible above the eastern horizon.

4. Set the program to the ending time of the observing session. What new Solar System objects, if any, are visible above the eastern horizon?

5. If possible, obtain for each Solar System object visible during the observing session the following data: rise and set times, magnitude, constellation in which it currently resides, current distance from Earth, and angular size as seen from Earth. Arrange these data in a table in your report.

6. Set the time to the middle of the observing session. Set the program to show the entire sky (this may be a mercator projection with a 180° field of view). Turn on deep sky objects with labels. List the deep sky objects (star clusters, nebulae, galaxies, etc.) near the meridian. Of these deep sky objects, which are brighter than magnitude +8?

7. Set the time to the middle of the observing session. Set the program to show the sky in a mercator projection. Turn on constellation lines and object labeling. Produce a hardcopy of this view.

8. Using the data gathered in the above steps, make a list of "good" objects (both Solar System and deep sky) for observing during the session. Typically, brighter objects are easier to find than dimmer ones. Celestial objects near the horizon appear distorted by the atmosphere and may be obscured by clouds, trees, or other objects on the ground. Finally, a wide range of objects makes observing more interesting.

Astronomy on the Internet

Computer networks provide astronomers with unprecedented opportunities to explore the final frontier, from remotely controlling observatory telescopes to retrieving the latest data from robot space probes to chatting with other astronomers. This lab is intended as an introduction to the rapidly evolving information super-highway.

The **Internet** is a world-wide network of millions of computer systems. Users of the Internet, often called "net surfers," range from research scientists to students to business persons to just about anyone wishing to share information electronically. There are a variety of ways to access information on the Internet, including electronic mail, newsgroups, IRC, telnet, File Transfer Protocol, Gopher, and World-Wide Web.

The first part of the Internet encountered by most people is **electronic mail** (or **e-mail**). E-mail allows users to communicate information and ideas across the world or next door with a minimum of effort and expense. Since most documents today are prepared on computers, it is often faster, less expensive, and environmentally conscious to transmit those documents via an electronic network than to print the documents on paper and have them delivered by traditional mail (or **snail-mail**). As with snail-mail, e-mail can not be delivered without a valid address. On local networks, an e-mail address may be as simple as the user's name. But on large networks, consisting of many computer systems, e-mail addresses must contain the system's address as well as the user's name. Thus, e-mail addresses often appear in the form,

> *user@computer.address*,

where the '@' symbol separates the user's name from the computer's address.

Every computer system on the Internet has an address, called an **IP number** (Internet Protocol number). An IP number is like a zip code and street address all rolled into one. It is represented by four integers, between 0 and 255, separated by periods, such as 123.4.100.15. IP numbers are hard to remember and do not provide much flexibility, so most people use domain names. A **domain name** consists of a series of words separated by periods, where each word refers to a link in the hierarchy of domains in which the computer system is located. The leftmost word in the domain name is the name of the host computer. It is followed by ever more general names. A computer system may be referred to by several domain names, which all correspond to the same IP number.

Domain names are either organizational or geographical. One of NASA's archive computers has the organizational domain name of *explorer.arc.nasa.gov*. This is a computer system, called *explorer*, that handles one of the archives (arc) at NASA, a government organization (gov). The rightmost name in the domain name is called the top-level domain and is the most general domain. Some top-level organizational domains are listed in Table 1.

Table 1: Some top-level organizational domain names.

Domain Name	Description
com	commercial organization
edu	educational institute
gov	government organization
mil	US military organization
org	nonprofit organization

The geographical domain name, *well.sf.ca.us*, refers to a computer system, called *The Well*, in San Francisco (sf), California (ca), USA (us). Some top-level geographical domains are listed in Table 2. Geographical domains are not unique across levels. For example, 'ca' represents California as well as Canada, but it only means California when the top-level domain is 'us'.

Table 2: Some top-level geographical domain names.

Domain Name	Description
au	Australia
ca	Canada
cl	Chile
dk	Denmark
fi	Finland
fr	France
de	Germany
it	Italy
jp	Japan
tw	Taiwan
uk	United Kingdom/Ireland
us	United States of America

After e-mail, one of the first stops of many net surfers is **Usenet**. Usenet is a network of over a hundred thousand computers that provides a sort of bulletin board of specialized topic areas, called **newsgroups**. People post notes (called articles) to these groups, and, hopefully, receive useful responses from others who read the articles in the group. There are newsgroups for a wide range of interests. The Usenet system uses a hierarchical naming system for the newsgroups. A name consists of a series of words separated by periods. For example, *sci.astro.amateur* is a newsgroup in which amateur astronomers post articles about their observations and discuss topics related to amateur astronomy. Many newsgroups regularly post **Frequently Asked Questions (FAQ)** lists. There is no need to post a question if it has already been answered in the group's FAQ. There is a wide variety of software for accessing the Usenet newsgroups, from simple text-based programs to programs with graphical user interfaces.

Another interesting part of the Internet is **Internet Relay Chat (IRC)**. IRC allows people to hold real-time conversations with lots of others with similar interests. IRC is divided into **channels**. Some channels are reserved for specific topics, others are generally available to anyone who **logs-into** (accesses) them. Users accessing a channel can transmit what they type on their computer terminals to the terminals of all or some of the other people logged into that channel. Announcements for special astronomy chat sessions are often announced in the sci.astro Usenet newsgroup. These astronomy sessions permit professional and amateur astronomers to chat about each other's observations, the latest data from space probes, and other topics of astronomy and space.

Telnet is a class of programs that allow one computer (called a client) to log into another computer (called a host) through the Internet and act as though it were a terminal directly connected to that host. Some Internet resources are only available through telnet. A host system is accessed by giving the telnet program the host's IP number or domain name. Once connected, the host computer will prompt the remote user for **login** information, such as user ID and password, if required.

Many host systems are set up to allow users to **download** (copy files from the host) and **upload** (copy files to the host). One of the oldest and most common methods for accomplishing this is through **File Transfer Protocol (FTP)**. Using an FTP program is similar to using a telnet program. The user must give the FTP program the address of the host system. Once a connection is made, the host computer provides login instructions.

Some hosts provide access to some of their data through "anonymous" FTP, in which the requested user ID is "anonymous" and the password is the electronic mail address of the remote user. Once logged in, the user may move to different directories, perform directory listings, and download and upload files. Some FTP commands are listed in Table 3. Figure 39-1 gives a sample FTP session in which an image is downloaded.

One of the difficulties of using the Internet is in finding specific information in the overwhelming amount of data that is available. **Archie** is a project of the McGill University School of Computer Science that maintains a database of files available through anonymous FTP. This FTP database is then distributed to other computer systems (called archie servers), where Internet users may query the database for the locations of files meeting specific characteristics.

Gopher is a simple, menu-driven program for accessing the Internet. Gopher was developed at the University of Minnesota in 1991 to provide students with easy access to campus information and has grown into one of the most popular means of accessing the Internet. "Gopherspace" is the name given to that part of the Internet accessible via gopher. There are many on-line libraries, universities, research labs, and even shops in gopherspace. Gopherspace can be searched using veronica. **Veronica**, which stands for "Very Easy Rodent-Oriented Net-wide Index to Computer Archives," is a database of menu items on the

```
ftp explorer.arc.nasa.gov  [This starts the FTP program and has it contact this site]
Name (explorer.arc.nasa.gov:userid): anonymous
["userid" is the ID of the person who initiated the FTP session.  This site allows anonymous
FTP access.]
331 Guest login ok, send your complete email address as password.
Password: userid@host.domain.name
[The user must enter his/her email address for the password.]
230-Welcome to explorer.arc.nasa.gov. The current time is Sun Jan  8 22:35:18 1995.
230 Guest login ok, access restrictions apply.
ftp>  [Whenever this prompt appears, the user can enter a command.]
ftp> ls  [The user enters the command to get a directory listing.]
200 PORT command successful.
150 Opening ASCII mode data connection for file list.
bin
etc
pub
dev
tmp
usr
cdrom
226 Transfer complete.
37 bytes received in 0.01 seconds (3.6 Kbytes/s)
[The host computer prints a listing of the current directory to the screen.  The amount of
data, transfer time, and transfer rate are reported.]
ftp> cd  pub/SPACE/GIF
[The user changes the directory that he/she is in.  A directory listing can be requested.]
250 CWD command successful.
ftp> binary
200 Type set to I.
[The user sets the data type to binary.  This must be done to successfully transfer binary
data, like images.  Text data should be transferred as ASCII.]
ftp> get mars.gif  [The user makes a request to get (download) the image called "mars.gif."]
200 PORT command successful.
150 Opening BINARY mode data connection for mars.gif (157528 bytes).
226 Transfer complete.
local: mars.gif remote: mars.gif
157528 bytes received in 30 seconds (5.1 Kbytes/s)
[The image was transferred as binary data to the user's computer.]
ftp> quit
221 Goodbye.  [The user disconnects from Explorer.]
```

Figure 1: This is a partial transcript of a FTP session in which an image is downloaded. The
user's keystrokes are printed in italics. Comments are made within square brackets.

Table 3: Some FTP commands

Command	Description
ascii	switch to ASCII transfer (for text)
binary	switch to BINARY transfer (for non-text, like images and sounds)
cd *dirname*	change to directory *dirname*
dir	list contents of current directory with names, owners, permissions, and sizes
help	list FTP commands
get *filename*	copy specified file from host to local
ls	list contents of current directory by name only
put *filename*	copy specified file from local to host
mget *filename1 filename2* ...	copy multiple files from host
mput filename1 filename2 ...	copy multiple files to host
quit	close connection to host

gopher sites that can be reached from the gopher server at the University of Minnesota. Unlike archie, all veronica servers may not have the same information, so queries may need to be made to more than one veronica server.

The **World-Wide Web (WWW or W3 or, simply, the Web)** is an interlinking of the systems on the Internet by **HyperText Transfer Protocol (HTTP)**. Documents in webspace are written in the **HyperText Markup Language (HTML)**, which allows embedded images, sounds, and links to other WWW documents. **Web browsers** are programs that allow users to search "webspace." Most browsers take advantage of the Web's variety of resources and provide users with a friendly, graphical interface. The user can download data or jump to other documents simply by clicking the cursor on the appropriate icon or text.

Documents in webspace are addressed by **Uniform Resource Locators (URL's)**. A URL has the form,

resource-type://host.domain[:port]/path/filename,

where *resource-type* is the requested type of Internet connection (e.g., WWW, gopher, FTP, etc.), *host.domain* is the domain name of the host computer, sometimes a special *port* number is also required, *path* is the path of directories to the file, and *filename* is the name of the file. For example,

http://info.cern.ch/hypertext/DataSources/bySubject/Overview.html

refers to the hypertext document, *Overview.html*, in the */hypertext/ DataSources/ bySubject* directory of the system at *info.cern.ch* connected via the World-Wide Web (resource type http). Some common resource types are listed in Table 4. Notice that gopher, FTP, telnet, and other resources are also accessible through webspace. Hence, they are often listed by their URL's; for example,

ftp://explorer.arc.nasa.gov/pub/SPACE and *news:sci.astro.*

Note that the URL's of newsgroups lack the two forward slashes (//).

Table 4: Some URL Resource Types

Resource Type	Description
file	a file on the local system or at an FTP site
http	a file on a WWW system
gopher	a file on a gopher system
WAIS	a file on a WAIS system
news	a Usenet newsgroup
telnet	a telnet service

Some observatories allow amateur astronomers to dial-in and control their telescopes from their home computers. For a flat fee, the user can direct the telescope to a particular area of the sky using special software. A charge-coupled camera system attached to the telescope captures the image and sends it to the user. These services give amateur astronomers access to equipment that is usually available only to professional astronomers.

The Internet is quickly evolving, so this text can not remain up-to-date nor can it list all available services. But, this laboratory exercise should get the student started exploring the Universe through resources available on computer networks. Table 5 lists some resources available on the Internet.

Table 5: Some Internet Resources

URL	Description
news:alt.binaries.pictures.astro	unencoded astronomy images
news:alt.sci.planetary	planetary sciences
news:sci.astro	general astronomy discussion
news:sci.astro.amateur	amateur astronomy discussion
news:sci.astro.hubble	Space Telescope observing schedules and info
news:sci.space.news	space-related news
news:sci.space.policy	space policy and government discussion
news:sci.space.science	space and planetary science discussion
ftp://ftl.jpl.nasa.gov/pub	Jet Propulsion Laboratory
ftp://explorer.arc.nasa.gov/pub/SPACE	NASA Ames Archive
ftp://ftp.hq.nasa.gov/pub	NASA headquarters
ftp://ftp.stsci.edu/stsci/epa	Space Telescope Science Institute
gopher://spacelink.msfc.nasa.gov/	NASA SpaceLink gopher connection
gopher://gopher.tc.umn.edu	University of Minnesota Mother Gopher
http://aas.org/AAS-homepage.html	American Astronomical Society
http://cannon.sfsu.edu/~williams/ planetsearch/planetsearch.html	Information on extra solar planets
http://www.ari.net/nss	National Space Society (USA)
http://www.halebopp.com	Comet Hale-Bopp home page
http://www.jaxnet.com/~rcurry/nefas.html	Northeast Florida Astonomical Society
http://www.jpl.nasa.gov	Jet Propulsion Laboratory
http://www.kalmbach.com/astro/astronomy. html	Astronomy Magazine home page
http://www.mtwilson.edu	Mount Wilson Observatory
http://www.odysee.com.au	Southern Astronomical Society (Australia)
http://www.osf.hq.nasa.gov/shuttle/futsts. html	Future space shuttle missions
http://www.sel.noaa.gov/current_images	Current solar images
http://www.seti-inst.edu/	SETI Institute
http://www.skypub.com	Sky and Telescope Magazine home page
http://www.stsci.edu	Space Telescope Science Institute

Procedures

Apparatus
access to the Internet,
instructions for accessing the Internet, and
some astronomy Internet sites.

Note 1: Students will need at least one double-sided, high-density disk to complete this lab.

Note 2: The following sections of this lab are independent of each other and can be done in any order.

A. Usenet Newsgroups

1. Start the software needed to access the Usenet newsgroups.

2. List the names of the available astronomy newsgroups.

3. List the names of the available space newsgroups.

4. Choose an astronomy or space newsgroup and browse through its articles. Describe the newsgroup in your report.

5. If a current FAQ has been posted to the newsgroup, download it to your disk. In your report, note the filename under which the FAQ was saved.

B. File Transfer Protocol (FTP)

1. Acquire the domain name or IP number of an astronomy or space FTP site from the instructor.

2. Use the FTP software to connect to the site.

3. Download a text or image file from the site. Place this file on your disk, and use an appropriate program to view it. In your report, describe this file, give its filename, and note its location on the site. See Figure 39-1 for an example of an FTP session.

C. Telnet

1. Acquire the domain name or IP number of an astronomy or space telnet site from the instructor.

2. Telnet to the system and explore the available features.

3. Describe the features available at this telnet site. Also, provide a brief transcript of your telnet session in your report.

D. Gopher

1. Start the gopher software and select menu items to bring you to an astronomy or space site as per your instructor's directions.

2. Explore this gopher site. Describe its features and links in your report.

3. Download a text or image file using the gopher software. Save this file to your disk, and use an appropriate program to view it. In your report, describe this file, list its filename, and note the address of the gopher site from which it was obtained.

4. Access one of the links available from this site. Describe the new site in your report. Be sure to give the address and title of the new site. Repeat steps D.2 and D.3 for the new site.

E. World-Wide Web

1. Acquire the URL of a World-Wide Web site from your instructor.

2. Start the Web browsing software and open a connection to the site.

3. Explore the Web site. Describe its features and links in your report.

4. Download a text or image file from the site. Place this file on your disk, and use an appropriate program to view it. In your report, describe the file, give its filename, and note the URL of the site from which it was obtained.

5. Access one of the links available from this site. Describe the new site in your report. Be sure to give the URL and title of the new site. Repeat steps E.2 through E.4 for the new site.

Observatory Visit

Astronomical observatories are typically buildings that house large optical telescopes. The buildings and telescopes are usually designed and constructed for the specific site occupied by the observatory. This exercise considers observatories devoted to optical astronomy, but there are other types which can also provide unique opportunities for learning about astronomy. Such sites include observatories devoted to infrared, microwave, and radio astronomy, sites where robot space telescopes are controlled, and where neutrino detectors are located. Many observatories are associated with research institutions, and may not have facilities to accommodate visiting students.

Many observatories are located in remote sites, such as on top of mountains, far from city lights and air pollution. Desert conditions are preferable to avoid water vapor and clouds. NASA even has an observatory in a Lockheed C-141 airplane, the Kuiper Airborne Observatory, which can fly 13 km above Earth. This places it above 99% of the atmospheric water vapor and virtually all of Earth's weather systems. A telescope views the sky through an opening in the aircraft's side. Larger telescopes can be located on mountain tops. An excellent location is on Mauna Kea, an extinct volcano in Hawaii that reaches 4200 meters above sea-level. Mauna Kea has unusually good "seeing" conditions which result from a uniform flow of stable Pacific air over the mountain. As a result, several optical telescopes there typically achieve better than 1 second-of-arc resolution. Because of these excellent observing conditions, fourteen observatories populate this mountain top, and more are planned.

The latitude of an observatory is also an important consideration, for it determines what parts of the sky can be seen. Longitude does not matter since the Earth sequentially rotates observatories through all values of right ascension each day. The North Pole, for example, would be a poor site for an observatory devoted to general observation, even apart from the bad weather usually occurring there, because a telescope there would not be able to see any object south of the celestial equator. An observatory on the equator, on the other hand, would be able to see all of both the northern and southern celestial hemispheres. Traditionally, observatories were built in the northern hemispheres since the nations there industrialized first. Consequently, the northern skies have been more carefully studied than southern skies. More recent observatories have been built in the southern hemisphere, including those in Australia and Chile. Mauna Kea is about 20° north of the equator, so telescopes there can see most of the sky, excluding only the part within 20° of the southern celestial pole.

A traditional observatory houses its telescopes under a domed roof, and can open a gap in the dome which extends from the base to the top. The dome can rotate with the gap open, allowing the telescopes to scan the entire sky visible from that location. The dome is closed during the day to protect the telescopes from the Sun's rays and warm day-time air. The dome may also be closed during bad weather. The dome, even when open, provides protection from winds. Wind blowing over an unprotected telescope causes it to move and vibrate, producing blurred images.

The images can be blurred from vibrations resulting from the wind, movement of building elevators, and from local auto, truck and train traffic. Consequently, observatory telescopes are usually mounted on specially designed structures to avoid transmitting building and Earth vibrations to the telescopes. Large telescopes can be massive. For example, the Hale telescope at Mount Palomar has a mass of 500 tons. The supporting structures for such telescopes must be strong and must allow the telescopes to move and track objects in the sky. See Exercise 9, Telescopes II, for information on telescope mounts.

Optical telescopes are metal and glass devices built to exacting physical dimensions. If part of a telescope is warmed by the Sun or a current of warm air, thermal expansion of that part will distort the telescope and the images it produces. More time is required for massive parts to warm up or cool off than is required for other, less massive, parts. Hence, observatories are designed for operations that help minimize temperature fluctuations inside the domes. Failures can limit the telescope's usefulness. For example, when the observatory at Mt. Wilson was first used, multiple and distorted images of Jupiter were obtained on "first light," the first use of the telescope. These multiple images resulted from uneven heating of the mirror which inadvertently occurred during the dedication ceremony. It was not until the early morning hours that the temperature had equalized enough that the telescope finally gave a single sharp image.

Research observatories are air-conditioned, not for the comfort of those using the facilities, but to adjust and maintain the telescope's temperature to that expected for the outside air during the upcoming night's observing. This procedure minimizes thermal distortions. Not only is the temperature of the dome air adjusted, but in some newer observatories cooling coils are installed inside the floor to allow its temperature to be controlled, and to prevent the floor from radiating heat to the telescope. And heat from electronic equipment used in the dome is also carefully removed, and not just vented to ambient air inside the dome as is done in most other buildings.

The types of telescopes found in an observatory depend on the mission and age of the facility. Observatories located at universities and museums, and in or near cities, usually have teaching and public education as part of their missions. Such facilities include some that were constructed early this century and were at that time premier astronomy research facilities. Others include observatories that were constructed primarily for teaching and student research activities, and may include several identical telescopes housed in a large room covered by a flat roof, which can slide quickly out of the way for observing.

The design and operation of telescopes continues to evolve and to produce better instruments capable of providing more detailed images of distant objects in the night sky. Many new telescopes are both large and expensive. For example, the effective diameter of the objective mirror of each of the two Keck telescopes on Mauna Kea is 10 meters. These telescopes cost $100 million each. Astronomers rarely look thorough such instruments.

Instead, imaging is done with electronic devices (see Exercise 13, Electronic Imaging), in which the images are recorded digitally and displayed on monitors. This permits observers to be at remote sites, which allows observers to avoid a cold observatory, and also the oxygen deprivation and high altitude sickness which sometimes occur at high altitude observatories. This also reduces the time and expense that would be required for transportation to the remote observatory sites. The Keck telescopes on Mauna Kea can be operated from the Keck Center at Waimea Kamuela, a colorful village close to sea-level at the base of Mauna Kea.

Visiting an observatory provides opportunities to learn about many aspects of astronomy, including what it is like to work as an astronomer. Observatory etiquette is based on the idea that the research mission of the observatory is to be served before other missions. Observatory visitors should remember they are guests, and that astronomers working in the observatory may have waited for years to conduct the experiment they are now engaged in. If they fail to complete their work during their scheduled time, they will have to wait again. And it is possible that the opportunity to observe some particular astronomical event will never come again.

Procedures

A. Pre-Observatory Visit (as a class)

1. Discuss the directions to the observatory, meeting time and anticipated travel arrangements and plans. Discuss what clothing will be appropriate, and other personal information. Skirts should not be worn. Will hats, coats and gloves be needed? Will food and beverages be available or allowed? Will the number of toilets be adequate for the number of visitors? Review observatory etiquette. Will souvenirs, astronomical slides, photographs or postcards be available?

2. Review any available materials concerning the history and missions of the observatory, the types of research conducted, significant discoveries made there, and any public programs provided by the observatory. Describe the observatory latitude and altitude, and discuss the significance of that location.

3. Review the kinds of material that should be included in student reports about the visit, if required. Suggest topics and strategies for note taking.

B. At the Observatory

1. Follow the instructions of the instructor and observatory personnel. Be alert to what can be learned. Listen to the questions and comments of others.

2. Take notes and make sketches on the types of telescopes, mountings, and other equipment inside the dome. Record any information provided on the equipment's use, telescope magnification, resolution, etc.

C. Report

1. Describe the observatory visited, the type(s) and diameter(s) of the telescope(s), the type(s) of telescope mounts, and auxiliary equipment, such as spectroscopes, CCDs, etc.

2. Were you able to look through a telescope, or see images presented on a monitor? If so, describe how and sketch what was observed. What methods were used, photography or CCDs? What was the magnification? Note interesting or unusual features.

3. What are the advantages and disadvantages of a permanent observatory over portable telescopes? What are the advantages and disadvantages of an observatory being near a city versus in a remote site?

4. What were the responsibilities of the technician/observer/researcher which you met during the visit?

5. Make comments evaluating the observatory visit.

Planetarium Visit

The term **planetarium** was used in the 19th century to describe mechanical devices used to show how movements of the Sun, Moon, and planets created the views of these objects seen in Earth's sky. Seven objects that wander across the sky against the background of stars were known as "planets" to ancient astrologers. These were the Sun, Moon, and five of the planets, known today as Mercury, Venus, Mars, Jupiter, and Saturn. There are seven days in a week because seven "planets" were seen. The days are named in reverence to these "planets": **Sun**day, Monday or **Moon**day, and Saturday or **Saturn**day, are the most obvious. The term planetarium currently refers to a movie-theater like facility in which the movements of the Sun, Moon, planets, and stars can be projected on the inside of a large whitened dome. The first modern planetarium was built for the Deutsches Museum in Munich, Germany in 1923.

There are two basic systems used today for planetarium projectors: an optical–mechanical system and a computerized-vidicon system. In the optical–mechanical system, the main projector has one or two metal spheres containing powerful light sources that project light through holes arranged in their surfaces to present star patterns on the dome ceiling. The holes are of different sizes to allow different amounts of light through and to give the correct relative brightness of each star. There may be as many as 27,000 holes on these projectors. Rotating the projectors inside the planetarium dome can simulate the sky's movement. Additional and separate projectors create the relative motion of the Sun, Moon, and planets. Other projectors can also produce light patterns to simulate the band of the Milky Way and other deep sky objects. The control of the lamps, positions, and movements of all of these projectors requires the coordination of multiple controls, which are usually managed on computerized consoles.

A planetarium that employs a computerized-vidicon system uses computers to create star pattern graphics on bright monitors, called vidicon projectors, and project those patterns onto the dome. With the appropriate computer programs, this system allows more flexibility in creating different views. For example, the night sky as seen from the opposite side of the Milky Way galaxy can be shown. Other effects can also be produced, for example, flying around a mountain on Mars or through a valley on Venus. However, the "stars" produced by this system are not as bright or sharp, nor are they the correct colors, owing to current limitations in vidicon projectors.

Planetaria provide advantages for illustrating many lessons in basic astronomy. In planetaria, as in movie-theaters, the audience can lean back and relax in comfortable seats in a quiet, air conditioned and darkened room, while focusing their attentions on the "show" being presented. Often, even when the sky being projected mirrors what is outside, spectators inside notice details they do not notice outside. This is because of the many distractions typically present outside, because lighted pointers can be used to single out stars, and the names and outlines of constellations can be projected on the planetarium dome. Additionally, movements of the projected sky in planetaria can be speeded up to show years of movements in only a few minutes, or stopped and reversed. The audience can see how the sky appears from various other places on Earth, for example, from the north and south poles. The ecliptic, the meridian, and scales of coordinate systems, altitude/azimuth and right ascension/declination, can be shown. Finally, the "sky is always clear" inside a planetarium.

Modern planetaria are also like movie-theaters in that they usually have powerful, high quality sound systems, which can be used in multimedia presentations. Students who have never visited a planetarium will likely feel comfortable in them because of the movie-theater atmosphere. Theater etiquette is also appropriate for guiding individual behavior. Visitors should not leave the planetarium during a show, as light from outside may be let in, ruining the view and the night vision of others.

Procedures

A. Pre-Planetarium Visit (as a class)
1. Describe the directions to the planetarium, any admission fee required, meeting time, and travel arrangements. Indicate if students may invite others to join them. Distribute brochures from the planetarium, if available, and describe the types of programs the planetarium presents to the public.

2. Discuss the type of shows that will be presented. Explain what kind of report, if any, will be required from the students. Discuss strategies for taking notes. Review planetarium etiquette.

B. At the Planetarium
Follow the instructions of the instructor and planetarium personnel. Watch the shows, and take notes (on paper or mentally) on appropriate topics.

C. Report (if required)
1. Describe the planetarium visited, including the type of projector and any special features of the facility.

2. Describe the shows you watched and make comments evaluating the shows and presentations, and the overall value to you of the planetarium visit.

Radioactivity and Time

Radioactivity was discovered in 1896 when Antoine Bacquerel noticed that photographic film completely covered with black paper was exposed simply by being near rocks that contained uranium. Investigations revealed the "activity" that exposed the film traveled away from the minerals along radial lines. Hence, the name radioactivity was applied to describe this phenomenon. Three types of invisible rays, now known as **alpha**, **beta**, and **gamma rays**, were soon discovered. This unanticipated discovery followed the discovery of x-rays by only a few years. Gamma rays, like x-rays, are energetic particles of light called **photons**. Beta rays were later discovered to be electrons, and alpha rays were determined to be the nuclei of helium atoms. See Exercise 6, About Your Eyes, for an experiment with alpha rays. This exercise includes experiments in which students measure the half-lives of some radioactive elements. The **half-life** is the amount of time required for a radioactive element to lose half of its activity.

Radioactivity has provided so many new methods of studying nature, that it is hard to imagine where science would be without it. Prior to the discovery of radioactivity, it was believed that atoms were permanent and immutable. But naturally occurring radioactive isotopes provide scores of examples of atoms that disintegrate by themselves, producing new atoms of other elements, called **descendants**, many of which are themselves radioactive. Indeed, a basic lesson from Bacquerel's discovery is that the Earth is naturally radioactive. A partial list of radioactive isotopes is provided on the periodic table included in Exercise 33, Elements and Supernovae. Nine of the elements that occur naturally on Earth, from polonium to uranium, occur only in radioactive form, and result from the three decay series illustrated in Figure 42-1.

Three types of uranium atoms occur in the minerals of Earth, called **isotopes** of uranium, which differ in their masses and radioactive properties: uranium 234, uranium 235, and uranium 238. Other uranium isotopes have been made in nuclear physics laboratories. All have 92 protons in their nuclei, which is what makes them uranium, but each has a different number of neutrons, 142, 143, and 146 for the isotopes listed above. Notice that the sum of the number of protons and the number of neutrons in an atom is the atom's **isotope number**. The naturally occurring isotopes of uranium disintegrate by emitting alpha, beta, and gamma rays as they transmute through a series of descendants, illustrated in Figure 42-1. Some uranium atoms also split into two major parts in a process called spontaneous fissioning, which typically releases several neutrons.

Early studies indicated that the radioactive decay mode and half life of an atom does not depend upon the atom's environment, such as the temperature and pressure where the atom is or whether the atom is in a solid metal crystal, a liquid, or a gas. This led to the idea that radioactivity is a property of part of the atom that is remote from the atom's outer electrons. Later studies by Rutherford showed that the atomic nucleus is very small, and that radioactivity is a process of the nucleus.

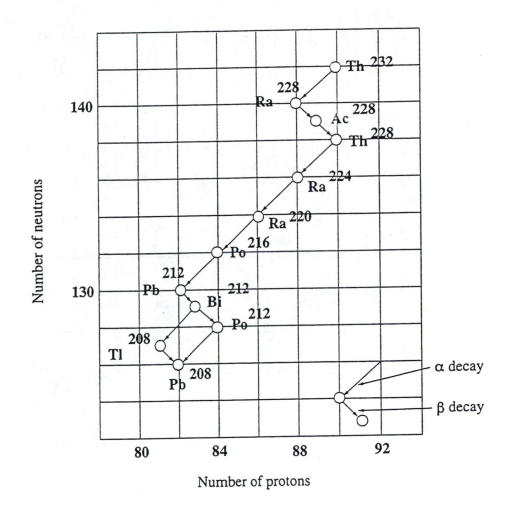

Figure 42-1.A: A plot of the decay series resulting from thorium 232, showing the neutron number versus the proton number of each of the descendants. The chemical symbols for these elements can be found in periodic tables, such as the one in Exercise 33, Elements and Supernovae. Note that an alpha decay produces a descendant with two fewer protons and neutrons, while a beta decay converts a neutron to a proton. For example, thorium 232 decays by emitting an alpha particle to become radium 228, which decays by beta emission to become actinium 228. Gamma rays, being photons, do not alter the identity or position of isotopes on this chart. Note that all the atomic mass numbers in this chart are all divisible by 4, so this series is called either the thorium series, or the 4n series. The value of n is different for each value of the atomic mass number. For example, n is 58 for thorium 232, and 57 for radium 228. Each atom of thorium 232 that completes this series produces 6 alpha particles (helium atoms) and one lead 208 atom.

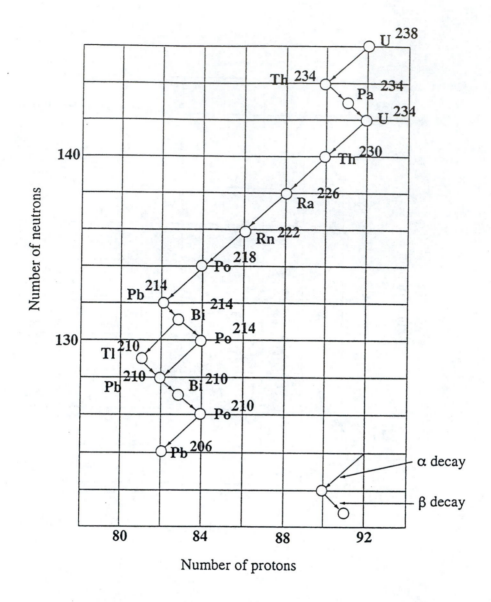

Figure 42-1.B: The decay of uranium 238 produces the descendants illustrated. Note that the atomic mass numbers are all equal to 4n + 2. For example, 238 is equal to 4 times 59 plus 2. Each atom of uranium 238 that completes this series produces 8 alpha particles and one lead 206 atom.

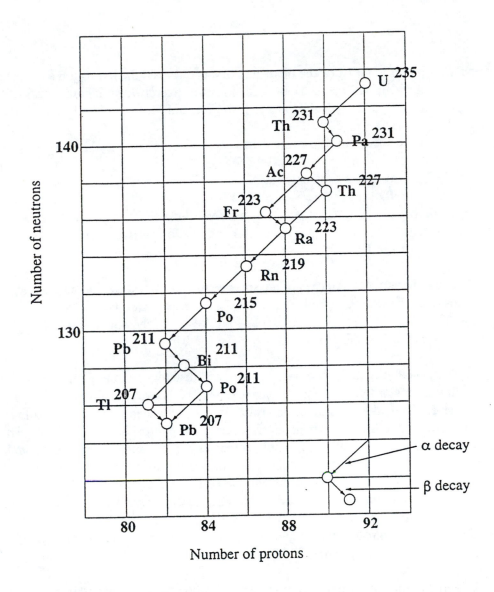

Figure 42-1.C: The decay of uranium 235 produces the descendants illustrated. Note that the atomic mass numbers are all equal to 4n + 3. Each atom of uranium 235 that completes this series produces 7 alpha particles and one lead 207 atom.

Example 1: Find the ratios of the diameters of a proton and the hydrogen atom, and the diameter of Earth and it's orbit. Data needed for this exercise are the diameters of the proton, about $3 \cdot 10^{-15}$ m, hydrogen atom, $1 \cdot 10^{-10}$ m, and Earth, 12,756 km, and two AU, about $3 \cdot 10^{11}$ m.

The ratio of the atom to proton diameters is,
$$(1 \cdot 10^{-10} \text{ m}) / (3 \cdot 10^{-15} \text{ m}) = 3 \cdot 10^4,$$
while 2 AU to Earth's diameter is,
$$(3 \cdot 10^{11} \text{ m}) / (13 \cdot 10^6 \text{ m}) = 2 \cdot 10^4.$$
It is, perhaps, a strange perspective that something as close together as the electron and proton in a hydrogen atom are, by comparison, more remote than the Sun is from Earth.

The radioactive properties of an isotope are determined by the nucleus, and these properties differ from those of all other isotopes. For example, the radioactive half-lives of uranium 238, uranium 235, and uranium 234 are, respectively, 4.47 billion years, 0.704 billion years, and 0.246 million years.

The probability that a radioactive atom will disintegrate in a given second is the same for all other atoms of the same isotope independent of the particular atom's age or environment. Thus, the disintegration probability is unchanged whether the atom is inside a rock on Earth or the Moon, or in an asteroid, or in the photosphere of the Sun. This means that individual radioactive atoms can not provide any information about their own history, but if they are frozen inside rocks, collectively they can provide information about the history of those rocks since they solidified. Determining the age of rocks by analysis of radioactive isotopes and their descendants is called **radioactive dating**, and such methods have been applied to many rocks from Earth and the Moon as well as to meteorites. This is one of the methods through which geologists and astronomers have come to believe that Earth and the Solar System are 4.6 billion years old.

Three definitions are useful in discussing radioactive isotopes: Avagrado's number, atomic mass unit, and mole. A **mole** is a specific number of atoms of a substance, called **Avagrado's number**, which is $6.02 \cdot 10^{26}$ atoms. A mole of hydrogen atoms has a mass of 1 kg, for example, while a mole of uranium has a mass of 238 kg. Ignoring errors of typically 1 percent, each atom has a mass in **atomic mass units** that is equal to its isotope number.

Example 2: Calculate the mass in kilograms of one atom of hydrogen 1.
An atom of hydrogen 1 has a mass of 1.0 kg$/6.02 \cdot 10^{26} = 1.7 \cdot 10^{-27}$ kg.

Example 3: Calculate the probability of decay per second for the naturally occurring uranium isotopes, and determine the number of atoms that would, on average, decay each second in one mole of each.
If half of a sample of atoms decays in a period of time T, the half-life, then the probability for an atom's decay per second turns out to be about $0.693/T$, and

the number of disintegrations per second in a sample of N atoms is this probability times N. The number 0.693 is approximately the natural logarithm of 2. For uranium 238 this probability is

$$0.693 / [(4.47 \cdot 10^9 \text{ years})(3.16 \cdot 10^7 \text{ sec/year})] = 4.9 \cdot 10^{-18} / \text{sec}.$$

So, in a mole there are

$$(4.9 \cdot 10^{-18})(6.02 \cdot 10^{26}) = 3.0 \cdot 10^9 \text{ disintegrations per second}.$$

In a mole of uranium 235 and 234, the equivalent values are $19 \cdot 10^9$ and $54 \cdot 10^{12}$, respectively.

Since the activity of a sample is proportional to the number of radioactive atoms in the sample, the activity will decrease as the number of radioactive atoms decreases. This produces a type of change called an **exponential decay**. Figure 42-2 illustrates such a decay. Although, mathematically, the exponential decay curve never reaches zero, one mole of radioactive atoms will on average decay to a single atom in about 89 half-lives. So, practically speaking, a radioactive isotope will have completely transmuted to its descendants in less than 100 half-lives.

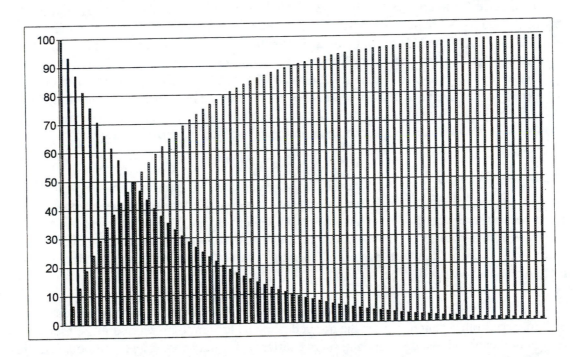

Figure 42-2: This graph shows an exponential decay as time increases (to the right), which is characteristic of radioactive isotopes. If a parent isotope decays to a non-radioactive descendant, then both the amount of the parent and the activity of a sample containing it will decrease in time as shown by the decreasing curve. The amount of the descendant will increase in time as shown by the other curve. Note that the sum of the two equals a constant of 100% all of the time.

Example 4: Assuming that equal numbers of uranium 238 and uranium 235 atoms were frozen inside a rock, what would be the ratio of the numbers of these atoms after 4.47 billion years?

The number of uranium 238 atoms would drop by half in 4.47 billion years, while the number of uranium 235 atoms would drop by half each time 0.704 billion years passes. The number of uranium 235 atoms left after 4.47 billion years is

$$(0.50)^{(4.47 / 0.704)} = 0.012 \text{ or } 1.2\%.$$

This means the abundance has fallen from 100/100 to 50/1.2 or about 98/2.

The term **relative abundance** identifies the percentage of atoms of an element that are a particular isotope, averaged for the materials of Earth. For example, the relative abundance of naturally occurring uranium isotopes are 99.27% for uranium 238, 0.720% for uranium 235, and 0.0055% for uranium 234. The relative abundance of uranium 235 is somewhat less than would be expected from the above example given that Earth is believed to be 4.6 billion years old. And the amount of uranium 234 is far greater than could be expected since 4.6 billion years is over 18,000 half-lives of uranium 234. But uranium 234 has continued to be created as a descendant of uranium 238, as can be seen in Figure 42-1.B. And uranium 235 and uranium 238 must have previously existed in the nebula from which the Solar System was formed before being incorporated into the Earth, and that time would also reduce the relative amount of uranium 235. Additionally, the abundances of these may never have been exactly equal. The method of dating minerals just described is called the **uranium-uranium** method, and although there are uncertainties in some cases, this method provides valuable information on the ages of rocks which correspond well with results from other dating methods.

If a rock froze containing thorium, but no lead or helium, then billions of years later the rock would contain fewer thorium atoms, and for each thorium atom that completed the decay series there would be one lead atom and 6 helium atoms. The age of this rock could therefore be determined from measuring the amount of lead and helium in the rock. Note that age here means "the time since solidification." Helium, being an inert gas, would escape molten material destined to become rocks.

The half-life of thorium 232 is 14 billion years. A rock must be crushed under carefully controlled conditions for the trapped lead and helium to be recovered. Procedures for doing this have been developed in several different laboratories. These methods, which were first developed using uranium, are called the **uranium-lead** and **uranium-helium** methods. Often all three methods, uranium-uranium, uranium-lead, and uranium-helium, can be applied to the same rocks. Other methods of radioactive dating, which apply to rocks billions of years old, and will not be discussed here, use potassium 40, rubidium 87, samarium 147, and rhenium 187. These isotopes have respective half-lives of 1.3 billion years, 48 billion years, 106 billion years, and 41 billion years. Collectively, these methods leave little doubt in the minds of geologists and astronomers that these methods are valid and that Earth and the Solar System are about 4.6 billion years old.

Example 5: Assuming that equal numbers of uranium 238 and 235 atoms were frozen inside a lead-free rock, what would be the ratio of the number of lead 207 atoms to the number of lead 206 atoms after 4.47 billion years? Ignore the decay time of all descendants.

From example 4, half of the uranium 238 atoms would have been transmuted to lead 206, while 98.8% of the uranium 235 atoms would have been transmuted to lead 207, so the ratio of lead isotopes would be 98.8/50.0 or 1.98. This example provides a basis for understanding why the ratio of lead isotopes from rocks which contain uranium differs from that found in rocks which contain no uranium.

Figure 42-1 shows three radioactive series which are identified as 4n, 4n+2, and 4n+3 series. Each is headed by a radioactive isotope with a half-life greater than half a billion years. One may wonder why there is no 4n+1 series. The answer to this question can be found by examining the half-lives of isotopes that would belong to a 4n+1 series. There are no long-lived isotopes that could make up such a series. Neptunium 237 is the isotope with the longest half-life, and its half-life is only 2.25 million years. It seems likely that early in the Solar System, there was a 4n+1 series headed by Neptunium 237. But with a 2.25 million year half-life, that series is now long since decayed.

Alpha rays were identified as helium atoms in 1909 by Rutherford and Royds, who trapped alpha rays in a glass tube that had two metal electrodes. When they created an electric arc through the tube and examined the resulting spectrum, they saw the spectral lines of helium.

The discovery of radioactivity had to await the development of photography, since there was previously no convenient method to detect it. Humans have no sense organs that can directly detect radioactivity. Subsequent developments in the field of electronics have made a number of detectors available for radioactivity, one of which will be used in this exercise. Understanding radioactivity and nuclear reactions has helped explain many things that would otherwise not be understandable. For example, how the interior of Earth, some 4.6 billion years after formation, can still be hot enough for convection to occur. This convection is responsible for earthquakes, Earth's magnetism, volcanism, and drifting continents. Calculations show that if it were not for the energy released by radioactivity, Earth's interior would have long since cooled. Many nuclear reactions release a large amount of energy, a million times the energy per reacting atom than is released in the most energetic chemical reactions, such as combustion. This helps to explain the origin of the energy powering the Sun and other stars.

Radioactivity also has many industrial and scientific uses, for example, to provide electric power for spacecraft operating in the dark. Robot spacecraft investigating the inner Solar System use photocells and sunlight to generate electric power for their operations. But in the outer Solar System, it is too dark for this to work. Radioactive materials carried on those spacecraft provide heat, which is used to generate electric power, making close-up views of the outer planets, their satellites, and rings possible.

A commercial minigenerator will be used in this exercise, which contains radioactive cesium 137 atoms that are chemically fixed inside the device. Cesium 137 has a half-life of 30 years, so measurements of the activity of the minigenerator at the beginning and end of the laboratory period should not show any significant change. Cesium 137 decays to an excited state of barium 137, called barium 137m. Since barium has different chemical properties, being a different element, it can be washed out of the minigenerator by passing a few drops of an acidic-salt water through it. The radioactive barium 137m atoms in this water will decay by gamma ray emission to become barium 137, which is not radioactive. The half-life of barium 137m is short enough that it will be seen to decay during the laboratory period. The experiment then is to determine by direct measurements the half-life of barium 137m. These measurements will be slightly complicated by the fact that radiations from natural sources will also be detected. This is called **background radiation** and it comes from the Earth, the laboratory building, and also from the cosmos. The latter radiation is reduced by the atmosphere at lower elevations, but can be eliminated only by going perhaps one kilometer underground.

Procedures

Apparatus
> cesium/barium minigenerator,
> Geiger tube with cable, planchet, and stand,
> sealed radioactive source,
> semi-log graph paper (4 cycles),
> scaler, timer, and connected power supply, and
> paper towels.

CAUTION 1: Do not allow the radioactive solution to come into contact with your body. If it is spilled, wipe it up with the paper towels provided, and place those towels in the provided container. No eating or drinking is allowed in this lab. Avoid touching yourself, and particularly avoid putting your fingers in your eyes, ears, nose, or mouth, as this may spread radioactive materials that you have inadvertently contacted. Wash your hands after cleaning up any spill, and before leaving the laboratory.

CAUTION 2: Do not touch the end window of the Geiger tube, as that may rupture the thin film window and also leave oil and salt deposits (finger prints) on it, which are difficult to remove. This window admits radiation into the Geiger tube which would not be able to penetrate the thicker side walls of the tube. The sensitive volume of such tubes typically starts a few millimeters behind the film window.

A. Calibration
1. Make sure the minigenerator is at least a meter away from your Geiger tube. Plug the power cord into a 115 VAC receptacle, and turn the scaler on. Set the clock to count for 30 second periods. Place the Geiger tube in its stand, and attach it to the scaler with its cable. Set the power supply to 900 V and turn it on. Push the reset and then the start button and observe that the counter occasionally detects radiation and adds one to its count..

2. Restart the counter and record the number of counts obtained in 30 seconds. Then push reset and start to count another 30 seconds. Repeat for four measurements of the background. Record each value.

3. Determine the mean number of counts in 30 seconds. It is normal for this background radiation to vary. Background counts will occur during all subsequent measurements and this mean background count may be subtracted. It is statistically important to subtract the background when the counting rate is low enough to be comparable to the background, but at times this can also produce negative counting rates.

B. Why it is called Radiation
1. Place your minigenerator (or another radiation source) close to the end of the Geiger tube with the film window, and record the number of counts obtained in 30 seconds. You will want to repeat this measurement at the end of the exercise, so note carefully the location

of the Geiger tube and radioactive source, so they can be returned as nearly as possible to the same positions.

2. Move the radiation source so that its distance from the Geiger tube window increases by 2 centimeters. Measure and record that distance, then operate the counter to determine and record the number of counts obtained in 30 seconds. Doubling the distance between the radiation source and the Geiger tube window, repeat the measurements and record the data.

3. Note that the count rate decreases as the minigenerator is moved farther from the detector. Why is that to be expected? The absorption of the air between the source and detector may be ignored here.

C. Determining Half-lives

1. In this step, you are to place a fresh solution containing barium 137m directly under your Geiger tube, and measure its activity versus time. A good procedure with this isotope is to count for 30 seconds starting every minute. You can tell when to start the 30 second counting period by watching the second hand of a wrist watch or wall clock.

> a. Practice the procedure of recording the counts, resetting the counter, starting the timer as the second hand passes say '12', recording the counts when the 30 second period is over, resetting the counter, and then starting the next count when the second hand passes '12' the next time (do not start the next 30 second period by waiting 30 seconds after the last such period ended, as that procedure will sequentially add any timing errors made in each period).

> b. Pass two or three drops of the acidic-salt solution through the minigenerator and onto a clean planchet. Carefully, but quickly, place the planchet under the Geiger tube, and begin recording its activity every minute, as just described.

> c. Continue counting and recording the data until you have obtained several values that are similar to the background values obtained previously.

2. Plot the counts, which are proportional to the sample activity, versus time on a semi-log plot, showing the time (number of minutes) on the horizontal, linear axis. Note that on 4-cycle semilog paper, if the bottom line is 1, the counts increase going up from 1, 2, 3, and so on to 9, and the next 1 is 10. The next 2 above 10 is 20, and so on to the next 1, which is 100, and the next 1 is 1000, and finally the 1 on the top line is 10,000.

Your data should display a straight line which slopes down, but then levels off and becomes equal to the background. There may be scatter in the background, as cosmic rays typically come in showers. The half-life can be obtained by drawing the best straight line through the sloping data, and finding how many minutes are required for the activity to drop by half, say from 1000 to 500, or from 200 to 100. Record your half-life value.

3. Return your radiation source and Geiger tube to the positions they had in B.1. above, and repeat that measurement.

Can you determine the half-life for cesium 137 from these measurements? What procedures would you recommend for determining the half-life of a sample which was expected to be one thousand years? One billion years? (Hints: What other information would need to be known about the samples? Also see example 2 above.)

Classifying Galaxies

The word "galaxy," having been used in English since the fourteenth century, is as old as this language. Galaxy was derived from the French, Greek, and Latin words for milk. To pre-industrial people, lacking bright lights, the Milky Way, a band of diffuse light stretching across the dark sky, would have been as familiar as the planets and the Moon. This band of stars completely encircles Earth. It is the disk of our Galaxy seen from the inside (but not from the center). With unaided eyes one other galaxy can be seen in the northern sky, the Andromeda nebulae, as it was called before its true nature was known. It is a faint fuzzy patch in the region of the sky containing the stars of the constellation Andromeda. Two galaxies, the Large and Small Magellanic Clouds, can be seen with unaided eyes in the southern sky from south of about 10 degrees north latitude. The Magellanic Clouds were described by sailors from Magellan's voyage, and are now known to be small irregular galaxies, that orbit the Milky Way. Our Sun and all of the individual stars that can be seen with unaided eyes from Earth are part of the Milky Way.

Immanuel Kant (1724-1804) speculated that the faint patches of light, which improved telescopes revealed in large numbers, were "...island universes - in other words, Milky Ways ..." Astronomers of the eighteenth century identified celestial objects as being either stellar or non-stellar, with the second category including gaseous nebulae, planetary nebulae, hazy star clusters, and faint lens-shaped formations. These structures were listed in catalogues according to brightness, appearance, and position in the sky. See Exercise 5, The Messier List. Unlike the gaseous nebulae that populate the disk of the Milky Way, the objects referred to by Kant are found in all directions of the sky except where obscuring dust clouds in the Milky Way intervene. The weight of astronomical opinion rejected Kant's speculation for over 150 years.

By 1910, nearly 15,000 nebulae had been catalogued. Some were correctly identified as star clusters and others as gaseous nebulae (such as the Great Nebula in Orion). Most, however, remained unexplained. If they were nearby, with distances comparable to those of observable stars, they would have to be luminous clouds of gas within our Galaxy. If, on the other hand, they were very remote, far beyond the foreground stars of the Galaxy, they would be systems containing billions of stars. They would have to be island universes, galaxies, as Kant had described them.

Edwin Hubble published a paper in 1929, "A Spiral Nebula as a Stellar System," which showed that the fuzzy patch in the constellation Andromeda is a system of numerous individual stars, star clusters, and dust clouds. Hubble's work expanded our knowledge of the Universe far beyond the Milky Way, and galaxy research has been an important part of astronomy ever since. The Andromeda galaxy, as it is now known, is 2 million light-years away, and is the nearest spiral galaxy to the Milky Way.

Galaxies, like stars, exist in clusters. It is now known that the Milky Way is part of a local cluster of about 30 galaxies called the **Local Group**. The Local Group lies on the outskirts of a much larger cluster called the **Local Supercluster**. A nearby cluster, the Virgo cluster of galaxies, is prominent in the sky because of its relative nearness. It lies in the Virgo and Coma

Berenices constellations, at a distance of about 65 million light-years, and contains many bright objects, some with Messier numbers and hundreds with NGC (*New General Catalogue*) numbers. This cluster has an angular diameter of about 10 degrees.

Example 1: How far is it across the Virgo cluster of galaxies?
The distance across the Virgo cluster is the product of 65 million light-years and the tangent of 10°, which is about 11 million light-years.

High quality astrophotographs of this Virgo Cluster show thousands of individual galaxies that are ideal for studying the various types of galaxies. There are several advantages of studying clusters of galaxies, the members of clusters are all at about the same distance, so relative luminosities and diameters can be easily compared for the cluster members. Examples of galaxies in collision can also often be found in galaxy clusters. Reproductions of original negatives of the Virgo and other galaxy clusters may be obtained from the Palomar Sky Survey, the Royal Observatory in Edinbourgh, and found in the *Hubble Atlas of Galaxies*.

Edwin Hubble and other observers made thousands of photographs of galaxies. Hubble found that almost all galaxies photographed could be classified into the small number of types identified on the tuning fork diagram illustrated in Figure 43-1. Hubble's system lists three basic categories: **elliptical galaxies**, **spiral galaxies**, and **irregular galaxies**. The spirals are divided into two groups, **normal** and **barred**. The elliptical galaxies, and both normal and barred spiral galaxies, are subdivided further, as illustrated in the figure, and as discussed below.

Spiral galaxies are associated in the public's mind with galaxies because their curving arms and dust clouds make spectacular pictures. Photographs of the Andromeda, the Whirlpool, and the Sombrero galaxies are often reproduced because of their beauty. The arms form a **disk** that extends out from a central or **nuclear bulge** of stars, often called the **nucleus**, which is brightest at its center. The arms require longer exposures to be seen in astrophotographs, which usually causes the nucleus to be overexposed. The disk and central nucleus of a spiral galaxy is enclosed in a larger spherical **halo**, which includes many globular clusters of stars. The halo is in turn surrounded by a dark **corona**, that has a radius of perhaps several times the radius of the disk. The corona is the location of the hypothetical dark mass, which can not be seen on photographs, and so will not be considered further here.

Spiral galaxies are classified by the shape of their nuclei and their arms. Spirals with arms that appear tightly wound are classified as Sa galaxies, while those with arms that appear more loosely wound are called Sc galaxies. Sb spirals are between these classes. Spirals with loosely wound arms also have smaller central bulges. The arms in barred spirals originate at the ends of of a bar running through the galaxy's nucleus. Some barred spirals have tightly wound arms, while other barred spirals have loosely wound arms, which are classified into Sba, Sbb, and Sbc categories. The Milky Way is probably an Sb galaxy. The disks of spiral galaxies are filled with clouds of dust and gas. The dust in the disk of the Milky Way may make more than half of the stars in our Galaxy invisible even in the largest telescopes. This dust prevents us from seeing (in visible light) more than about one-third of the distance to the galaxy center, which, from Earth, is

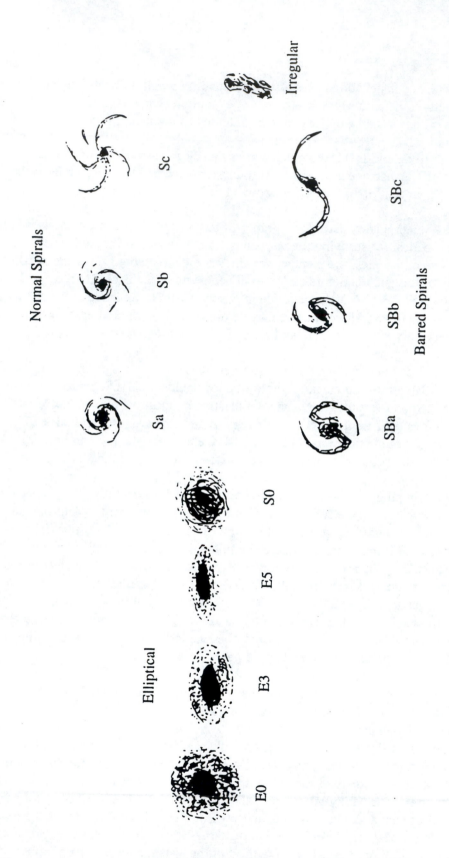

Figure 43-1. Hubble's Tuning Fork diagram for classifying galaxies.

in the direction of the constellation of Sagittarius. If the gas in the atmosphere of Earth was as dusty as the gas of the galaxy disk, one would not be able to see objects more than 1 meter away.

Ellipticals are galaxies whose images have elliptical outlines, without spiral arms, and often appear rather smooth and featureless. Ellipticals vary in shape from spherical to flat, lens-shaped formations. Hubble identified these galaxies with the letter E and classified them according to their elliptical shape using numbers from zero through seven. Spherical galaxies are classified as E0, almost spherical as E1, and galaxies with maximum elongation as E7. Elliptical galaxies contain relatively little dust and gas. Some giant ellipticals contain as many as a hundred times more stars than our Galaxy. M87 is an example of such a giant galaxy. Dwarf ellipticals are the most numerous type of galaxy, and typically contain only a few million stars. In some photographs of nearby dwarf galaxies, for example, Leo I which is about 1 million light years away, one can see through the center of the galaxy.

Since the formation of new stars occurs in regions which contain dust and gas, elliptical galaxies are composed primarily of old stars. The arms of the spiral and irregular galaxies not only show dust and gas clouds, but also contain hot blue stars, often within dust clouds, which is consistent with recent and ongoing star formation. See Exercise 32, Hertzsprung-Russell Diagram.

Still another class of galaxies intermediate between the elliptical and spiral galaxies are the S0 galaxies. Superficially, they appear like ellipticals, but they contain too much dust to be ellipticals, yet are too smooth to be spirals. Some show faint disks. The range of galaxies that exist seem to form a continuous sequence of galaxy types between the ellipticals and the spirals. Some galaxies are close enough to interact gravitationally and distort each others shapes. Irregular galaxies lack the organized appearance found in ellipticals and spirals. Given the estimated tens of billions of galaxies that exist in the observable Universe, it is remarkable that so many can be classified into so few categories.

The actual distribution of galaxies among the different types listed here is unknown and currently unknowable, as different imaging techniques skew the data in different ways. An early survey of prominent galaxies by Hubble netted 80% spirals. More recent surveys have shown spirals to be a smaller percentage. William Hartmann recently suggested that the distribution is "more like" 50% dwarf ellipticals, 5% giant ellipticals, 20% spirals, and 25% irregulars. Since dwarf ellipticals and irregulars are smaller and dimmer, they would likely be under counted in images of distant clusters. The Local Group of galaxies contains 3 spirals, one of which is the Milky Way, several irregulars including the Magellanic Clouds, and 12 or so dwarf ellipticals. Eight of the dwarf ellipticals and the two Magellanic Clouds are satellite galaxies of the Milky Way. This grouping has been referred to as Snow White and the seven dwarfs, even though the number of dwarfs is not seven.

Procedure

Apparatus

reproductions of photographs or negatives showing the Virgo cluster of galaxies, or other images of galaxies, hand magnifier, rectangular grid, and pads of small stick-on notes.

Caution: The photographs and negatives used in this exercise may be distorted and damaged by writing on them, or even by pressing on them with a pen, even if no ink is transferred. Identify and mark galaxies by using the grid overlay, and stick-on notes. If comments are to be made on the stick-on notes, write such comments before placing it on the negative.

A. Classifying Galaxies

1. Use a magnifier to examine the negatives or photographs showing galaxies. First scan the document to observe the range of objects shown without any attempt to make records. You need to learn to see details in these objects. Look for both similarities and differences, and think about how they could be classified as elliptical, spiral, barred spiral, or irregular.

Also notice that there may be galaxies close together, or even with overlapping images. As with binary and double stars, they may be gravitationally bound and interacting, or they may just appear to be together because they are in the same direction from Earth. Distortions can usually be seen in strongly interacting galaxies.

2. To avoid counting galaxies more than once, use the grids provided and divide the image areas into sections. Proceed systematically and use stick-on notes to mark your progress through the sections. Apply the classification system for the galaxies that you find. Classify each galaxy by considering its overall shape, prominence of its central bulge relative to its disk, and the structure of any arms. Record this information in a table.

B. The Relative Numbers of Galaxy Types

1. Calculate the percentage of each major type of galaxy (elliptical, normal spiral, barred spiral, and irregular) for the cluster. Were all types found? If not, can you account for those missing?

2. Would you expect the ratios of the different types of galaxies to differ for a more distant cluster where only the brightest galaxies are seen? Explain.

3. Could any objects in your classes represent objects in other classes seen from different angles?